CNC Prog

Volume 1 - 6

S. K. Sinha
Department of Mechanical Engineering
Indian Institute of Technology
BHU, Varanasi (INDIA)

CreateSpace, Charleston SC

Vol. 1: CNC Programming Skills: Program Entry and Editing on Fanuc Machines

Vol. 2: CNC Programming Skills: Understanding G73 on a Fanuc Lathe

Vol. 3: CNC Programming skills: Live Tool Drilling Cycles on a Fanuc Lathe

Vol. 4: CNC Programming Skills: Understanding Offsets on Fanuc Machines

Vol. 5: CNC Programming Skills: Understanding G32, G34, G76 and G92 on a Fanuc Lathe

Vol. 6: CNC Programming Skills: Understanding G71 and G72 on a Fanuc Lathe

Printed by CreateSpace,
An amazon.com Company

CNC Books by the Same Author

1. CNC Programming Using Fanuc Custom Macro B, published by **McGraw Hill, USA** http://www.amazon.com/dp/0071713328 (Also available in Chinese version published by **Science Press, Beijing** http://www.amazon.com/dp/7030311329)

2. CNC Programming Skills: Program Entry and Editing on Fanuc Machines (Vol. 1), Kindle edition published on amazon.com http://www.amazon.com/dp/B00R7JG5WO

3. CNC Programming Skills: Understanding G73 on a Fanuc Lathe (Vol. 2), Kindle edition published on amazon.com http://www.amazon.com/dp/B00SSFWH5A

4. CNC Programming skills: Live Tool Drilling Cycles on a Fanuc Lathe (Vol. 3), Kindle edition published on amazon.com http://www.amazon.com/dp/B00TGP74Z4

5. CNC Programming Skills: Understanding Offsets on Fanuc Machines (Vol. 4), Kindle edition published on amazon.com http://www.amazon.com/dp/B00VD9FQT2

6. CNC Programming Skills: Understanding G32, G34, G76 and G92 on a Fanuc Lathe (Vol. 5), Kindle edition published on amazon.com http://www.amazon.com/dp/B01C7T04OI

7. CNC Programming Skills: Understanding G71 and G72 on a Fanuc Lathe (Vol. 6), Kindle edition published on amazon.com http://www.amazon.com/dp/B01FOPN3RW

TABLE OF CONTENTS

VOLUME 1:

Program Entry and Editing
on
Fanuc Machines

PREFACE

This is the first book in the series *CNC Programming Skills*.

Do you know how to insert a part of a program into another program at the desired location? Background editing?? Using PCMCIA card??? Or, maybe, a simple task such as replacing G02 by G03 in the whole file????

When it comes to manual program entry on the machine, or searching / deleting / editing / copying / moving / inserting an existing program residing in the control memory or the PCMCIA card, most people resort to trial and error method. While they might be able to accomplish what they desire, the right approach would save a lot of their precious time. If this is exactly what you want, this book is for you. The information contained herein is concise, yet complete and exhaustive. The best part is that you can enjoy the convenience of having the wealth of useful information on editing techniques in a pocket-size book which you can always carry with you! You would often need to refer to it because it is not possible to memorize all the steps which are many a time too complex and devoid of common logic, so as to make the correct guess.

The procedure for program entry / editing on a Fanuc 0i control is explained after the definition of some key words, which applies to all i-series control versions (Modal C).

Happy editing ...

MDI PANEL

The keyboard which is an integral part of the LCD screen is called MDI (manual data input) panel. This is supplied by Fanuc and has a standard configuration for a particular control version / model. Fanuc 0i standard MDI has 24 address / numeric keys, 6 function keys (POS, PROG, OFS/SET, SYSTEM, MESSAGE, and CSTM/GR keys), 3 edit keys (ALTER, INSERT and DELETE keys), one SHIFT key, one CAN key, one INPUT key, one HELP key, one RESET key, 2 page change keys and four (left / right / up / down) arrow keys. The MDI panel is used for typing / editing a program and also for entering data for changing the software settings (such as setting system parameters for the overtravel limits, etc.) of the machine.

One can type a new program or edit an existing program directly on the MDI (Manual data input) panel of the machine in the EDIT mode. The control saves whatever we type, the moment we press the INSERT key on the MDI panel (except in *background editing* mode, described later). Similarly, if we delete a word(s) or a block(s), it is gone forever. A deleted file also cannot be recovered. There is no *undo* key or a *recycle bin*. So, if the backup of the old file is also desired, it has to be copied into a new file and then edited. The programs are executed in the automatic (AUTO) mode which is also referred to as the memory (MEM) mode. Small programs, which need not be saved for a future use, can also be executed in the manual data input (MDI) mode. Manual movement of the tool, say, for the purpose of offset setting, can be done in JOG / Handle mode. EDIT mode is only for the purpose of editing.

MOP PANEL

This panel (machine operator's panel), which does not have a standard configuration, is for selecting machine modes (EDIT, AUTO, REMOTE, MDI, JOG, INC, HANDLE, REFERENCE and TECAH modes, all of which may not be available on all the machines) and for hardware control of the machine such as coolant ON / OFF, spindle ON / OFF, feed drives ON / OFF, tailstock forward / retract, tool change, various overrides, program execution start / stop, feed hold, single block execution, block skip, optional stop, emergency stop etc. The panel also has some key switches for edit lock, machine lock, auxiliary (MST) lock, external / internal chucking, emergency override etc.

Keeping in view the capability of a machine (all the control features are not implemented in all the machines), the MOP panel is designed by the machine tool builder (MTB) who makes the hardware part of the machine and installs on it the control supplied by Fanuc (or some other company). The control of the machine is a complete package of the electronic parts as well as compatible servo motors / encoders with feedback circuits, servo amplifiers, cables etc. It is interesting to note that the machine control unit (MCU), i.e., the box which houses all the PCB's of the control is just about the size of a standard dictionary, irrespective of the size of the machine!

SOFT KEYS

Just below the LCD screen, there are 7 buttons, which have context based (as defined by the function keys on the MDI panel and subsequent selection of soft keys) multiple functions. In a particular application, the meaning of each button is displayed just above it. Since the functions of these bottoms vary and are software dependent, these are called soft keys. The extreme left button (◄) and the extreme right button (►) are for *return menu* and *next menu* (remaining options which could not be displayed due to lack of space), respectively. In most cases, OPRT (operation selection key) is indicated for the second button from the right. This button opens the menu for the possible operations on the selection made by the remaining four buttons (which are called chapter selection or context selection soft keys). In many cases, the chapter selection keys have a tree structure, i.e., selection of one chapter may open up several chapters, some of which may again have several chapters, and so on.

ENTERING A NEW PROGRAM

Assuming that the new program number is 1234, carry out the following steps:

- Select the EDIT mode on the MOP panel.
- Press the PROG key on the MDI panel.
- Type O1234 and press the INSERT key on the MDI panel.

If a program with number 1234 already exists, an error message would come. In such a case, either delete the old program or use an unused number. If no program with this number exists, then a blank program editing screen will appear with O1234 displayed in the first line and % in the second (last) line. Start typing the program, using the shift key, wherever required. Some of the keys, which are used for both characters and numbers, are context sensitive, i.e., if a number is required to be entered, the keys will behave as number keys. In a different context, these become character keys. For macro words, abbreviations can be used. For example, it is sufficient to type WH instead of WHILE. The program will insert and display the complete word only.

After typing each complete word (e.g., N100, G01, M03, X12.5, S1000 etc.), press the INSERT key. At the end of a block, press EOB (end of block) key, followed by INSERT key. This will insert a semicolon, after which typing starts from the next line. Insert EOB after O1234 also, to change the line. Note that it is not necessary to start typing the program from the second line, but it "looks" better that way. Insertion is done *after* the highlighted word. It is permitted to insert multiple words in one step – just type them sequentially, and press INSERT after typing the last word. In fact, multiple blocks (separated by EOB, i.e., semicolons) also can be inserted in one step, in the same manner.

While specifying the distances along different axes, be careful that a parameter setting may cause the *integer* values (i.e., distances in whole numbers) to be interpreted in the least input increment of the machine, i.e., as multiples of one micron (in mm mode). For example, X10 may be interpreted as 10 microns along X-axis. So, it is a safe practice to use X10.0 (or simply, X10.), if 10 mm is implied. The least input increment in inch mode is

0.0001 inch. So, in inch mode, X10 may be interpreted as 0.0010 inch. The least input increment is the mechanical resolution of the machine – the minimum distance by which the tool can be made to move along an axis, from its present position. In fact, a two dimensional (three dimensional on a 3-axis machine) uniform grid with the grid spacing equal to the least input increment can be thought to exist in the work envelope of the machine. The tool can only move to the discrete grid points – it cannot stop at an intermediate position. This happens because the control of the machine is *digital*, not analog.

For the purpose of editing or for simply displaying the desired blocks / words of the program, move the cursor using left / right / up / down arrow keys and page up / page down keys, as needed. The position of the cursor is indicated by highlighting the word at that position. The *complete* word is highlighted and the cursor moves word by word (by left / right arrow keys). This means that any editing operation is possible only on complete words – you cannot edit a part of a word; edit the whole word. In the discussion that follows, "bringing the cursor to a word" and "highlighting a word" – both mean the same thing.

Use CAN (cancel) key to delete one character at a time (backwards) *while typing*. DELETE key deletes the whole of the highlighted word. If a word is to be replaced by a new word, highlight the old word, type the new word and press the ALTER key. The program number (O-word in the first line of the program) and the end of file character (% character in the last line) cannot be deleted or altered.

For deleting a complete block (each line of a program is called a *block* in CNC language), place the cursor on the first word of the block, press EOB, followed by DELETE keys. Instead of placing the cursor on the first word, if it is placed on some other word, then all the words of that block except those lying to the left of the highlighted word would be deleted. It is also possible to delete multiple blocks, provided the last block to be deleted has a sequence number (say, N01000) – instead of EOB, type N01000 and press DELETE.

It is also possible to search a word (which consists of a *letter address* followed by *data*, e.g., S1200) by typing it *fully* or typing its *address only*, and pressing the soft key SRH↓ or SRH↑ for forward / backward search. For example S1200 cannot be searched by typing S1, but it can be searched by both S1200 and S. Also, leading zeroes at the left of a data are not ignored in a search operation. For example, S0100 cannot be searched by typing S100. Type either S0100 or just S. Typing just the address is very convenient if, for example, you want to examine the spindle speeds (S-word) specified in the whole program, one by one. For this, type S and press SRH↓ repeatedly, till you reach the end of the program where the search ends.

Comments, if any, must be enclosed between left and right parentheses, i.e., "(" and ")" characters which are called *control out* and *control in* characters, respectively, and the INSERT key should be pressed *after* the end of the complete comment.

If the MOP has a *block skip switch*, insert the slash character, /, in the *beginning* of a block to skip it optionally (depending on the ON / OFF setting of the block skip switch on the MOP, *at the time of reading the program*), during the execution of the program. For multiple block skip switches (if provided), insert /1, /2, /3 etc. (a maximum of nine block skip switches may be available), corresponding to the respective switch numbers. Some controls (including Fanuc 0i) allow the slash symbol even in the middle of a block (mid block skip) for the purpose of skipping the part of the block lying to the right of the slash, but it might cause confusion because the slash character is also used for the division operation in a macro statement (i.e., a statement in terms of *variables*). If the purpose is *division*, then it must be enclosed in square brackets. For example, in #101 = [#100/2], the slash means division, but in #101 = #100/2, the slash means mid block skip (if the block skip switch is ON), which sets #101 = #100. However, it is better to use conditional (IF _ GOTO _) or unconditional (GOTO _) branching instead of using block skip. Block skip was originally designed for machines not having macro programming option.

Use arrow keys and page up / page down keys to move the cursor wherever desired. The left / right arrows move the cursor backward / forward, one word at a time. The up / down arrow keys move the cursor to the first word of the previous / next block. Page up / page down keys move the cursor to the first word of the previous / next page. In all cases, the movement will be continuous if you hold down the keys. The RESET key places the cursor on the first word of the program.

The MDI panel usually has just one key for the left bracket and another one for the right bracket. While typing, whether the bracket will appear as a square bracket or a parenthesis, depends on a parameter setting. If the machine has Fanuc 0i or a similar control, set parameter 3204#0 to 0 for a square bracket, and 1 for a parenthesis. The first bit (from right) of an 8-bit parameter is referred to as bit #0 (the eight bits are designated as #0 through #7, starting from the right). Obviously, it would not be possible to use square brackets as well as insert comments both *at the same time*, with this parameter setting. If some comments are desired to be inserted in a macro program, first type / edit the program using square brackets, wherever required. Then change the parameter setting to get parentheses, for the purpose of inserting comments. After inserting comments, if again some corrections in the program, involving square brackets, are needed, another change in parameter 3204 would be required.

There is, however, a way to use both types of brackets, with the same parameter setting. For this, set parameter 3204#2 to 1 (3204#0 should remain 0. Status of the other six bits of this parameter does not affect this feature). Such a parameter setting displays an extended character set, as soft keys, in the EDIT mode, displaying "(", ")" and "@". With this setting, if square brackets are needed, use the bracket keys on the MDI panel, and if parentheses are needed, do the following (in EDIT mode):

• Press PROG on MDI panel (Press PROG again if the current program is not displayed).
• Press OPRT soft key.
• Press the next menu key (▶) twice.
• Press C-EXT soft key.

After this, soft keys for left parenthesis, right parenthesis and @ (which can be used in the comments inserted in a program) will appear which can be used as and when required, for editing in EDIT mode. However, a change in display screen will make these soft keys disappear. If these are again needed, the process to display these will have to be repeated.

EDITING / EXECUTING AN EXISTING PROGRAM

Assuming that the existing program number is 1234, carry out the following steps:

- Select the EDIT or AUTO mode on the MOP panel.
- Press the PROG key on the MDI panel.
- Type O1234 (or simply 1234) and press the O SRH (program number search) soft key. If O SRH does not appear, press the return menu soft key (◄).

   Program number 1234 will appear in the program screen, and O1234 will be displayed at the top right corner of the screen. If the specified program number does not exist, "DATA NOT FOUND" will be displayed. DIR soft key will display a list of defined programs. Note that though the program can be searched in EDIT as well as AUTO mode, editing is possible only in EDIT mode, and execution (i.e., machining the workpiece) in AUTO mode. *Background editing* (discussed later), however, is possible in AUTO mode also. For executing a program, i.e., for starting the machining, first set the proper conditions for machining (such as offset setting, hydraulic ON, tailstock EXTEND, releasing axis inhibit interlock etc.), bring the cursor to the first block of the program, and then press CYCLE START on the MOP.

DELETING A PROGRAM

Assuming that the existing program number 1234 is to be deleted, carry out the following steps:

- Select the EDIT mode on the MOP panel.
- Press the PROG key on the MDI panel.

- Type O1234 ("O" has to be typed; simply 1234 will not do.) and press the DELETE key on the MDI panel.

It is also possible to delete all programs within a defined range, in one step. For example, if all the programs including and lying between program numbers 1000 and 2000 are to be deleted, then in the third step, type O1000,O2000 and press the DELETE key. For deleting all the programs, type O-9999 and press the DELETE key.

COPYING AN ENTIRE PROGRAM

Assuming that the existing program number 1234 is to be copied to create a *new* program number 2345, carry out the following steps:

- Select the EDIT mode on the MOP panel.
- Press the PROG key on the MDI panel.
- Type O1234 (or simply 1234) and press the O SRH soft key.
- Press the soft key OPRT. If OPRT does not appear, press the return menu soft key (◄).
- Press the next menu soft key (►).
- Press the soft key EX-EDT (extended edit), followed by COPY and ALL soft keys.
- Type 2345 and press INPUT on the MDI panel.
- Press the soft key EXEC (execute).

After this operation, we will have two programs O1234 and O2345 with the *same* contents. One of these may be selected for editing, and the other may be left unchanged.

COPYING A PART OF A PROGRAM

Assuming that a part of the existing program number 1234 is to be copied to create a *new* program number 2345, carry out the following steps:

- Select the EDIT mode on the MOP panel.
- Press the PROG key on the MDI panel.
- Type O1234 (or simply 1234) and press the O SRH soft key.

- Press the soft key OPRT. If OPRT does not appear, press the return menu soft key (◀).
- Press the next menu soft key (▶).
- Press the soft key EX-EDT, followed by COPY.
- Move the cursor to the start of the range to be copied and press the soft key CRSL~.
- Move the cursor to the end of the range to be copied and press the soft key ~CRSL or ~BTTM (in the latter case, the program up to the end is copied, regardless of the position of the cursor).
- Type 2345 and press INPUT on the MDI panel.
- Press the soft key EXEC.

After this operation, program number 1234 will remain *unaffected*, and the new program number 2345 will have the selected part of program number 1234.

MOVING A PART OF A PROGRAM

It is similar to the previous case, except that the part which is copied to create a new program is *deleted* from the original program:

- Select the EDIT mode on the MOP panel.
- Press the PROG key on the MDI panel.
- Type O1234 (or simply 1234) and press the O SRH soft key.
- Press the soft key OPRT. If OPRT does not appear, press the return menu soft key (◀).
- Press the next menu soft key (▶).
- Press the soft key EX-EDT, followed by MOVE.
- Move the cursor to the start of the range to be moved and press the soft key CRSL~.
- Move the cursor to the end of the range to be moved and press the soft key ~CRSL or ~BTTM (in the latter case, the program up to the end is moved, regardless of the position of the cursor).
- Type 2345 and press INPUT on the MDI panel.
- Press the soft key EXEC.

After this operation, the new program number 2345 will contain the selected part of program number 1234, and this part will be *deleted* from the original program number1234.

In all the copying and moving operations, if a program number is not input before pressing the EXEC soft key, a new program number 0000 is created. This program can be edited in the usual manner, but cannot be executed.

INSERTING A PROGRAM INTO ANOTHER PROGRAM

An existing program (say, program number 2345) can be inserted at an arbitrary position in the current program (say, program number 1234):

- Select the EDIT mode on the MOP panel.
- Press the PROG key on the MDI panel.
- Type O1234 (or simply 1234) and press the O SRH soft key.
- Press the soft key OPRT. If OPRT does not appear, press the return menu soft key (◄).
- Press the next menu soft key (►).
- Press the soft key EX-EDT, followed by MERGE.
- Move the cursor to the word, *before* which program number 2345 is to be inserted, and press the soft key ~'CRSL or ~BTTM' (in the latter case, the insertion is done at the end of the current program, regardless of the position of the cursor).
- Type 2345 and press INPUT on the MDI panel.
- Press the soft key EXEC.

After this operation, program number 2345 is inserted at the desired location in program number 1234, creating a *modified* program number 1234. Program number 2345 remains unaffected.

INSERTING A PART OF A PROGRAM INTO ANOTHER PROGRAM

There is no direct method for this. Use a combination of previous methods. For example, using the method of *copying a part of a program*, create a temporary program with some *unused* program number, and then, insert it at the desired location using the method of *inserting a program into another program*. Finally, delete the temporary program. Alternatively, do not specify any name for the intermediate program, in which case the copied program would be saved in program number 0000.

SEARCHING AND REPLACING A WORD(S) BY ANOTHER WORD(S)

Carry out the following steps to replace the words (say, X100 Y200) by other words (say, X200 Z100 F60) in program number, say, 1234:

- Select the EDIT mode on the MOP panel.
- Press the PROG key on the MDI panel.
- Type O1234 (or simply 1234) and press the O SRH soft key. If the program is already displayed, press RESET to bring the cursor to the first word of the program, if the search operation from the beginning of the program is desired; otherwise, searching will be done from the *current position* of the cursor, till the end of the program. It is always *downward* search.
- Press the soft key OPRT. If OPRT does not appear, press the return menu soft key (◀).
- Press the next menu soft key (▶).
- Press the soft key EX-EDT, followed by CHANGE.
- Type X100 Y200 and press the soft key BEFORE.
- Type X200 Z100 F60 and press the soft key AFTER. The *first* occurrence of X100 Y200, at or after the current position of the cursor, will be highlighted.
- Press the soft key EXEC to do the replacement at *all* occurrences, starting from the highlighted word, till the end of

the file. Any occurrence(s) before the highlighted word will remain unaffected. .

Press the soft key EX-SGL (execute single) to do the replacement at the highlighted word *only*. Thereafter, the next occurrence is highlighted.

Press the soft key SKIP to *skip* the replacement at the current occurrence and search for the next occurrence. This option can be repeatedly used to look for the occurrence at a desired location. At the proper location, press the soft key EX-SGL for replacement. Any combination of EX-SGL and SKIP, followed by EXEC (if replacement at all the subsequent occurrences is desired) can be used.

Restriction: Not more than 15 characters can be specified for old or new words.

EDITING OF CUSTOM MACROS

The methods are the same as those used for ordinary programs, except that the use of abbreviations (first two characters or more) is permitted for macros. For example, if WHILE is to be typed, for searching or inserting, it is permissible to type only WH or WHI etc. The program displays and searches complete words even if abbreviations are used while typing. Set parameter 3204 appropriately to be able to use both square brackets as well as parentheses, as explained earlier.

BACKGROUND EDITING

Editing a program while executing another program is called background editing. The methods of editing are the same as those for ordinary editing (which is called foreground editing), discussed earlier. Though a program cannot be executed in EDIT mode, it is permissible to change to EDIT mode once the program starts executing. Assume that while program number 1234 is executing, program number 2345 is to be edited in the background. The following steps are needed:

- Select the EDIT or AUTO mode on the MOP panel.
- Press the PROG key on the MDI panel.
- Type O1234 (or simply 1234) and press the O SRH soft key.
- Change to AUTO mode and press CYCLE START on the MOP, after setting proper conditions for machining. The program will start executing (Do take all the standard precautions for executing a program.).
- Press the PROG key on the MDI panel.
- Press the soft key OPRT. If OPRT does not appear, press the return menu soft key (◄).
- Press the soft key BG-EDT (background editing). The background editing screen will appear, displaying PROGRAM (BG-EDIT) at the top left corner of the screen.
- Type O2345 (or simply 2345) and press the O SRH soft key.
- Now edit the program in the usual manner. After the background editing is complete, press the soft key OPRT (if BG-END does not appear), followed by BG-END (background editing end).

Restriction: In background editing, no attempt should be made to make any change in the program being executed. Also, the program being executed cannot call (as a subprogram) the program being edited in the background. Note that alarms of foreground operation (execution, in this case) do not affect the background operation, and similarly, alarms of the background operation do not affect the foreground operation. To differentiate between the two categories of alarms, the term *P/S alarm* (position / setting alarm) is used for the foreground alarm, and *BP/S alarm* is used for the background alarm.

EDIT LOCK FUNCTION

The MOP has a key switch which can be activated to disable any editing operation on all the programs. Even without using this switch, it is possible to disable deletion or editing of program numbers 8000 to 8999 through a parameter setting. Another parameter disables deletion or editing of program numbers 9000 to 9999. Additionally, it is also possible to password protect program numbers 9000 to 9999, which has an overriding effect on the parameter setting for this group of programs (in fact, the parameter cannot be changed to allow editing, if password protection is in effect). The password is set in a system parameter. With this protection enabled (i.e., a password provided), it is not possible to edit or delete these programs without providing the correct password (which is called *key word*), in another system parameter. Usually, edit lock function is used for protecting *tested* custom macros in which no change is needed, and which must be protected from inadvertent editing / deletion. However, care must be taken in using this feature, because if you forget the password (the password setting parameter does not display the set password), editing would be possible only after *memory all clear* operation! Finally, if memory all clear is really required then press RESET and DELETE simultaneously during power ON.

FLASH MEMORY CARD (PCMCIA CARD) AS AN EXTERNAL MEMORY DEVICE

Fanuc 0i control uses a compact flash memory card as an external memory device for storing part programs and CNC set up information. It is also referred to as PCMCIA card because it is connected to the PCMCIA port of the MCU. PCMCIA port for a computer has been standardized by *Personal Computer Memory Card International Association*. Interestingly, the lengthy acronym gave rise to the humorous expansion, "People Cannot Memorize Computer Industry Acronyms." PCMCIA card, which is now commonly called PC card, is about the size of a credit card. The memory capacity typically varies from 32 MB to 8 GB. There are three different sizes, varying in thickness: Type I is 3.3 mm thick, Type II is 5.0 mm thick and Type III is 10.5 mm thick. Recently, Toshiba introduced Type IV of 16 mm thickness, but this has not been officially standardized or sanctioned by the PCMCIA. All types are 85.6 mm long and 54.0 mm wide, along with the adapter for the PCMCIA port. An adapter is needed to connect the 50 pin memory card to the 68 pin PCMCIA port. Handle carefully so as to protect the pins. Also, do not apply any pressure at the centre of the memory card. Hold it by its edges.

Fanuc 0i accepts Type I flash memory card. Most machines do not accept cards with over 512 MB capacity. By setting parameter number 0020 to 4, files on the memory card, inserted into the PCMCIA slot which is located at the left of the LCD screen, can be accessed. Alternatively, press the OFS/SET key on the MDI twice (or press OFS/SET followed by SETING soft key), and edit to make I/O CHANNEL 4.

All the files stored on the memory card have sequential *file numbers* and specific *file names*. Though file numbers are always numeric (e.g., 0012), file names can have text characters also (e.g., O1234, TESTPROG etc.). File numbers, starting from 1 and in an increment of 1, are automatically assigned to every new file saved on the memory card. If a file on the card is deleted, the *subsequent files numbers are updated* (such that new file number = old file number + 1) to fill the created gap. Some possible operations on a memory card are described next.

DISPLAYING THE DIRECTORY OF STORED FILES ON THE MEMORY CARD

- Select the EDIT mode on the MOP panel.
- Press the PROG key on the MDI panel.
- Press the next menu soft key (▶).
- Press the soft key CARD. The directory will appear on the screen, with the heading DIRECTORY (M-CARD). The first two columns show file numbers and file names, respectively. Use page up / page down keys to see additional pages, if any. Comments, if any, in the first block of the files can be seen by pressing DIR+ soft key. For example, if the first block of the program is
O1234 (TENSILE SPECIMEN);
then (TENSILE SPECIMEN) will appear in the COMMENT column of the card directory.

SEARCHING FOR A FILE STORED IN THE MEMORY CARD

Files can be searched by the assigned *file numbers*. So, you must know which file number contains which file. For example, file number 12 may contain program number 1234. Carry out the following steps to search for file number 12:

- Select the EDIT mode on the MOP panel.
- Press the PROG key on the MDI panel.
- Press the next menu soft key (▶).
- Press the soft key CARD.
- Press the soft key OPRT.
- Press the soft key F SRH.
- Type 12 and press the soft key F SET.
- Press the soft key EXEC.

      The file number 12, if found, will be displayed at the top of the directory.

COPYING A FILE STORED IN THE MEMORY CARD USING FILE NUMBER

The following operation will read file number 12 from the memory card and will save it as program number 1234 in the CNC memory:

- Select the EDIT mode on the MOP panel.
- Press the PROG key on the MDI panel.
- Press the next menu soft key (▶).
- Press the soft key CARD.
- Press the soft key OPRT.
- Press the soft key F READ (file read).
- Type 12 and press the soft key F SET.
- Type 1234 and press the soft key O SET.
  If this step is skipped, the saved program number would be 0001, and if a program with this number already exists then an alarm would come and copying will not be done.
- Press the soft key EXEC.

While the program is being copied, the character string INPUT blinks at the lower right corner of the screen. Copying may take several seconds, depending on the size of the file.

Note that file number 12 need not be a part program file. It can be, for example, a parameter file, which will be saved in a format similar to a *notepad* file, as program number 1234. If program number 1234 already exists in the memory of the CNC, then it may or may not be overwritten depending on a parameter setting. While working with the memory card, if it is required to see / edit / delete the directory / programs saved in the control memory, press PRGRM soft key. If PRGRM is not displayed, press the return menu soft key (◄). For coming back to the card directory, repeat the initial steps (PROG → next menu → CARD). Repeat the last three steps to copy more files.

COPYING A FILE STORED IN THE MEMORY CARD USING FILE NAME

The following operation will read the file with name TESTPROG from the memory card and will save it as program number 1234 in the CNC memory:

- Select the EDIT mode on the MOP panel.
- Press the PROG key on the MDI panel.
- Press the next menu soft key (►).
- Press the soft key CARD.
- Press the soft key OPRT.
- Press the soft key N READ (name read).
- Type TESTPROG and press the soft key F NAME.
  The complete file name along with extension, if any, is required to be typed.
- Type 1234 and press the soft key O SET.
- Press the soft key EXEC.

Repeat the last three steps to copy more files. Finally, press the CAN soft key to cancel the current mode and go to the previous menu.

WRITING A FILE ON THE MEMORY CARD

The following operation will save program number 1234 in the memory card, with the name TESTPROG:

- Select the EDIT mode on the MOP panel.
- Press the PROG key on the MDI panel.
- Press the next menu soft key (▶).
- Press the soft key CARD.
- Press the soft key OPRT.
- Press the soft key PUNCH.
- Type 1234 and press the soft key O SET.
- Type TESTPROG and press the soft key F NAME.
- Press the soft key EXEC.

While the file is being copied on the memory card, the character string OUTPUT blinks at the lower right corner of the screen. Copying may take several seconds, depending on the size of the file being copied.

If a file with file name TESTPROG already exists in the memory card, it may be overwritten unconditionally or a message confirming the overwriting may be displayed, depending on a parameter setting. In case of such a warning message, press the EXEC soft key to overwrite, and CAN soft key to cancel writing. However, system information such as PMC ladder is always overwritten unconditionally. The copied file is automatically assigned the highest existing file number plus one. The comment, if any, with the O-word (i.e., in the first block of the program) will be displayed in the COMMENT column of the card directory.

To write all programs, type -9999 as the program number. In this case, if file name is not specified, all the programs are saved in file name PROGRAM.ALL on the memory card. A file name can have up to 8 characters, and an extension up to 3 characters (XXXXXXXX.XXX). Repeat the last three steps to copy more files. Finally, press the CAN soft key, to cancel the copying mode and go to the previous menu.

DELETING A FILE ON THE MEMORY CARD

Carry out the following steps to delete file number 12 (say) on the memory card:

- Select the EDIT mode on the MOP panel.
- Press the PROG key on the MDI panel.
- Press the next menu soft key (▶).
- Press the soft key CARD.
- Press the soft key OPRT.
- Press the soft key DELETE.
- Type 12 and press the soft key F SET.
- Press the soft key EXEC.

If more files are to be deleted, repeat the last two steps. Finally, press the CAN soft key to end the delete mode and go to the previous menu.

When file number 12 is deleted, this file number is assigned to the next file, and all the *subsequent* files are renumbered (new file number = old file number - 1). The numbers of the first 11 files (in this case) remain unchanged. Files on a memory card can only be deleted by file numbers; file names cannot be used.

VOLUME 2:

Understanding G73
on a
Fanuc Lathe

PREFACE

This is the second e-book in the series, "CNC Programming Skills." The first book which was published in December 2014 dealt with the program entry and editing procedures on Fanuc machines. Its successful launch encouraged the author to bring out another book on a topic which is not clearly understood by most CNC programmers. And, that is the canned cycle G73 on a Fanuc lathe. Most people do not use it even when it has got a perfect application, just because its intricacies are not properly understood.

How do you machine a job with a non-monotonic variation in surface profile, i.e., with radial or axial undercuts, if Type II G71 / G72 cycles are not available? May use a CAM software if you have access to it. That is perfectly okay. But, how do you machine a workpiece obtained by casting or forging with only slightly oversized dimensions, say, about 2 mm? Actually, this has got no simple solution if G73 is not used. G73 is designed mainly for such applications only. Read on and find out yourself. Manual part programming would never be the same again!

APPLICATIONS OF THE PATTERN REPEATING CYCLE, G73

The most efficient turning and facing cycles are G71 and G72, respectively. These, however, suffer from the limitation that they cannot make undercuts, unless type II cycles are available. For example, the job shown in Fig. 1, which has a radial undercut, cannot be produced by type I G71. Such jobs can be easily made by the pattern repeating cycle, G73.

It is possible to make both radial and axial undercuts by G73. Virtually any shape can be obtained by this cycle, provided suitable tools are available which obviate the problem of interference between the tool shank and the job.

Note, however, that even if a job has an undercut, it should first be machined using G71 / G72, leaving the undercut (assuming G71 / G72 type II cycles are not available). Thereafter, G73 should be used for making the undercut (only). This is because G73 is a very inefficient cycle compared to G71 / G72 (unless a cast / forged workpiece is to be machined, discussed next).

Another, and in fact, the most important use of G73 is for machining a cast / forged workpiece which is only slightly over-size by, say, 2-3 mm, in radial and / or axial directions. One cannot use G70 because removing 2-3 mm material in one pass may not be permissible. G71 or G72 would be too inefficient because the tool would cut in air most of the time. One can, of course, use a CAM software where the initial shape may be appropriately defined, but that also requires additional modeling time as well as software expertise. G73, on the other hand, can very quickly machine it to the final size in just a few passes (depending on the chosen depth of cut in one pass and the amount of over-size).

G73 can also be used for internal machining. Internal machining on the negative X side is exactly same as the external machining described here. However, if the internal machining is required to be done on the positive X side, the radial relief and the radial finishing allowance (defined later) would both be negative. Most machines are designed for doing internal machining on the positive X side.

Some restrictions, common to G71 and G72, apply to G73 also: In the blocks in which G71, G72 and G73 are commanded and between the block numbers specified by *P* and *Q*, M98 and M99 cannot be commanded. Also, these cycles cannot be executed in the DNC mode (because the program is fed to the machine line by line).

TOOLPATH IN G73

The toolpath in G73 is explained here with reference to a job with only radial undercut, requiring only radial relief and no axial relief. (Radial relief and axial relief are explained in detail later.) With a non-zero axial relief, the subsequent passes shift axially also. For jobs with only axial undercuts, zero radial relief and suitable axial relief is selected. A slightly over-size workpiece, made by, say, casting or forging, requires both radial relief and axial relief.

In this cycle, machining is done in several passes, all of which are equi-spaced and parallel (offset) to the defined surface profile. The first pass is farthest from the defined profile, and the subsequent passes gradually approach the defined profile, as shown in Fig. 1. The radial / axial displacement between the first pass and the last pass is called *radial / axial relief* (axial relief is zero in Fig. 1). The last pass exactly follows the defined surface profile (i.e., without any displacement), over which we have no control, except that if finishing allowances are specified, the last pass gets shifted radially by half of *X* finishing allowance (usually, *X* finishing allowance is on diameter) and axially by *Z* finishing allowance, with respect to the defined profile. All other passes remain parallel to the shifted profile (the last pass). Thus, the first pass gets displaced radially / axially from the defined profile by the specified radial / axial relief amounts plus finishing allowances.

Note the following in Fig. 1:

1. The position of the tool at the time of calling G73 automatically becomes the start point of G73. The sequence of motions is numbered. Feed motions are shown with continuous lines and rapid motions with dashed lines.

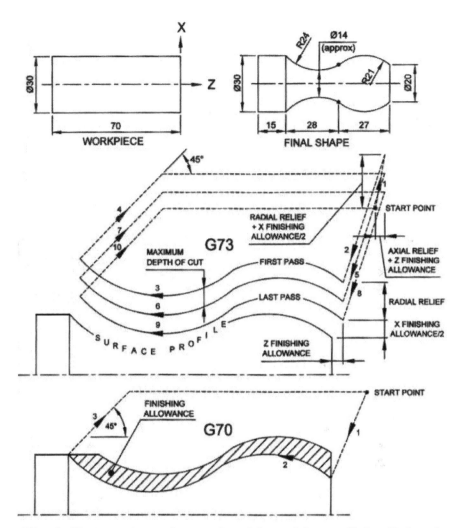

Fig. 1: Pattern repeating cycle with positive radial relief and zero axial relief

2. The retraction at the end of a cutting pass is rapid and shows dogleg effect (because initially both the axis motors start moving with rapid rate, with one motor stopping when the required distance in the corresponding axial direction is covered; the other motor continues moving till the desired end point is reached. The resulting angled motion is referred to as *dog-leg effect* because of its resemblance with the back leg of a dog).

43

3. The motion towards the start of a cutting pass is executed at rapid rate or feed rate depending on whether G00 or G01 is programmed in the *P*-block of the profile definition. The rapid motion with G00 may or may not show dog-leg effect depending on a parameter setting of the machine. This parameter, however, has no effect on the dog-leg effect during retraction. All these statements are valid for G70 also.

4. The last cutting pass of G73 is along the defined profile, shifted by radial and axial finishing allowances.

5. The first cutting pass of G73 is shifted with respect to the last cutting pass by the specified radial and axial relief amounts in the radial and axial directions, respectively (see Fig. 2 also). The associated approach and retraction motions are also shifted in a similar manner. All intermediate passes are evenly placed between the first pass and the last pass.

6. The direction of shift in all cases (i.e., due to relief amounts as well as due to finishing allowances) is along positive or negative axis direction, depending on whether the specified value has a positive or a negative sign.

7. Both G73 and G70 allow both *X*-word and *Z*-word in the *P*-block of the profile definition, and allow an angled approach motion (with or without dogleg effect, as stated earlier) towards the start point of a cutting pass (dogleg effect is not shown in the figure).

RADIAL RELIEF AND AXIAL RELIEF

The radial or the axial relief, if positive, shifts the first pass (as well as the intermediate passes) in the respective positive coordinate direction.

For a job with only radial undercut, such as the one in Fig. 1, the axial relief is kept zero (so that there is no axial shift in subsequent passes), and the radial relief is generally so chosen that the first pass just touches the workpiece at the *Z* location where the undercut has the minimum diameter. Though this is a safe practice, such a choice means that the actual machining would start only from the second pass. For such an arrangement, the radial relief is made equal to the depth of radial undercut (radial relief is a radius value). Similarly, for a job with only axial undercut, the radial relief is kept zero (so that there is no radial shift in subsequent passes), and the axial relief is set equal to the depth of the axial undercut, which would, likewise, start cutting material from the second pass.

Though it is possible to specify non-zero values for both radial relief and axial relief at the same time, only one is required to be non-zero (the other must be zero), in the case of only radial or only axial undercut. With a non-zero radial relief, the subsequent cutting passes shift radially, whereas a non-zero axial relief makes them shift in the axial direction, towards the defined profile. When both the relief values are non-zero, the shift is at some angle determined by the ratio of the two relief values. The shift is always towards the defined profile. Refer to Fig. 2 which explains different cases of relief values.

The relief amounts and the finishing allowances can be chosen to be negative also, to suit the given geometry and the orientation of the tool. For example, the radial relief as well as the radial finishing allowance both must be negative in the case of internal turning on the positive X side, because the first pass would need to be displaced in the negative X direction with respect to the defined profile, with the subsequent passes gradually shifting upward to match the defined profile.

Radial relief and axial relief (U and W, respectively, in the first block of G73) are calculated in the following manner:

U = Depth of radial undercut = (The largest diameter of radial undercut − The smallest diameter of radial undercut) / 2

With this value of U, the tool during the first pass will just touch the workpiece at the Z location where the surface profile has minimum diameter (maximum diameter, in case of internal machining), if there is no axial shift in the subsequent passes.

In the present example, the minimum diameter of undercut can be seen to be 14 mm (approx.), whereas the largest diameter of undercut is 30 mm. Thus,

$U = (30 - 14) / 2 = 8$

$W = 0$ (because axial shift in the subsequent passes is not required)

It is also possible to give undercuts on the face of a job. In such a case, the toolpath, which is again always parallel to the desired profile, is made to shift axially towards the face of the job from its extreme right position. The total axial displacement between the first pass and the last pass is the axial relief amount. The radial relief amount is, obviously, zero, since there is no radial displacement in the toolpath. The calculations are done in the following manner:

$U = 0$ (because radial shift in subsequent passes is not required)

W = The maximum depth of undercut from the face of the job

U and W calculated in this manner will cause the first pass of the tool to just touch the surface of the workpiece without removing any material. If machining is desired even in the first pass, the maximum permissible depth of cut should be deducted from the U and W values calculated here. The number of passes, R, (explained in the next section) also should be calculated using the new values for U and W.

A lower value of U and W than calculated can be used to complete an incomplete G73 cycle which might have been previously interrupted in the middle of execution due to some reason such as power failure. R will have to be recalculated for the new value of U or W. Then, the cycle will start machining from the point the execution was previously terminated. For example, if $U = 10$ and $R = 21$ (assuming the maximum depth of cut to be 0.5 mm), and the cycle is terminated after executing 10-11 passes, the new values for U and R would be 5 and 11, respectively, for the next G73 cycle.

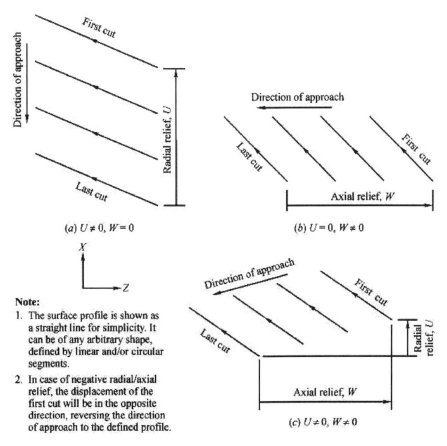

Fig. 2: Effect of positive radial / axial relief in G73

NUMBER OF PASSES

In this cycle, the depth of cut is specified in an indirect manner. Instead of depth of cut, the number of passes is specified, which is calculated on the basis of the permissible maximum depth of cut in a pass. All the passes are equi-spaced, the spacing being the maximum depth of cut in each pass (the depth of cut is, obviously, non-uniform in any pass, while machining an undercut). The spacing between the adjacent passes can be increased / decreased by decreasing / increasing the number of passes. Since the total displacement (i.e., radial / axial relief) is equally divided among all

47

the passes, the required number of passes (R) can be calculated for the desired spacing as

$R^* = (U$ or W / Maximum depth of cut$) + 1$

$R = R^*$, if R^* is a whole number

$R =$ Next whole number, if R^* is not a whole number

In the present example of non-zero U (= 8) and zero W, if we choose the maximum depth of cut to be 0.5 mm,

$R^* = 8 / 0.5 + 1 = 17$

Hence, $R = 17$.

FINISHING ALLOWANCES

It is possible to specify some finishing allowance also, to be later removed by the finishing cycle, G70. However, as explained below, X and Z finishing allowances should be zero for axial and radial undercuts, respectively. The effect of the signs of finishing allowances on the defined profile is same as that in G71 / G72; positive values shift the defined profile in the respective positive coordinate directions, and negative values in the negative directions.

In this example, both X and Z finishing allowances have been used. However, in case of radial undercut, the Z finishing allowance should be zero, because, otherwise, it will shift the final pass of G73 to the right (assuming a positive allowance) by the specified amount, which may result in overcutting of the right side of the undercut, as shown in Fig. 3 (b). Similarly, in case of axial undercuts, X finishing allowance should be zero to avoid overcutting, as shown in Fig. 4 (a).

Both X and Z finishing allowances may be specified only if there is no undercut on the profile. Such a case may arise when we want to machine a slightly over-size part. In this case, machining can be very conveniently and efficiently done by G73 where both X and Z finishing allowances may be specified for accuracy. The problem of overcutting due to finishing allowances does not arise in case of G71 or G72 type I cycles because undercuts are not included while defining the profiles for these cycles.

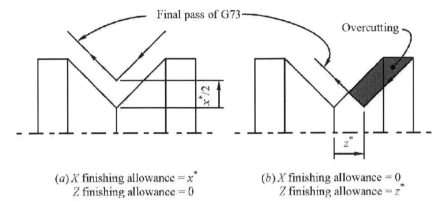

(a) X finishing allowance $= x^*$
Z finishing allowance $= 0$

(b) X finishing allowance $= 0$
Z finishing allowance $= z^*$

Fig. 3: Effect of positive X and Z finishing allowances on a radial undercut

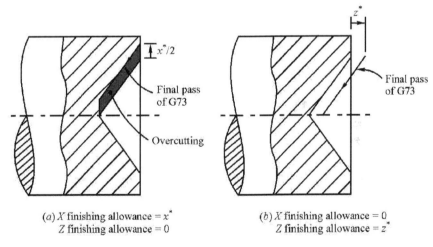

(a) X finishing allowance $= x^*$
Z finishing allowance $= 0$

(b) X finishing allowance $= 0$
Z finishing allowance $= z^*$

Fig. 4: Effect of positive X and Z finishing allowances on an axial undercut

FEEDRATE

G73 uses the feedrate specified in its second block. The feedrate as well as other modal data such as rpm (S-word) and tool offset (T-word) mentioned while defining the surface profile are ignored. These are used by G70, as in the case of G71 / G72. For the purpose of avoiding any confusion, it is recommended that only profile be defined between P and Q blocks. If different values for feed etc. are needed with G70, these may be explicitly commanded in the G70 block itself.

DEFINING THE PROFILE

While defining the profile (which can be defined in any direction which would determine the direction of machining, unlike the profiles for G71 / G72 which need to be defined in the proper directions), the first point (in P block) should always have a G00 address instead of G01; otherwise, not only G70, but G73 also will slow down because the tool would approach the start point of all passes at feedrate which should be avoided as it involves no machining (See also Note-3 corresponding to Fig. 1). The P block may have both X and Z words, unlike that in Type-1 G71 / G72. Only G00, G01, G02 and G03 can be used for defining the profile.

THE PROGRAM FOR THE FIG. 1 JOB

This is an example of a job with only radial undercut. The program for a job with only axial undercut can be written in a similar manner. Line numbers are optional except P-block and Q-block line numbers.

O0011; (Program number 11)

G21 G97 G98; (mm mode, constant RPM, and feed in mm/min selected)

G54; (Workpiece coordinate system selected)

G28 U0 W0; (Tool goes to its Home position)

T0909; (Tool no. 9 with offset no. 9 selected)

M03 S1200; (CW RPM = 1200 starts)

G00 X32 Z2; (Rapid positioning for G73, at the user's start point)

G73 U8 W0 R17; (U is the radial relief amount, W is the axial relief amount, and R is the total number of passes to be executed by G73)

G73 P1 Q2 U.1 W.1 F60; (The surface profile is defined between line numbers specified with P and Q. P is the starting line number, Q is the ending line number, U is the finishing allowance in the X direction (on diameter), W is the finishing allowance in the Z direction, and F is the feedrate)

N1 G00 X20 Z0; (The first point on the profile)

G03 X20 Z–27 R21;

N2 G02 X30 Z–55 R24; (The last point on the profile. If the profile is defined from left to right, the direction of passes will reverse)

M05; (Spindle stops)

G28 U0 W0; (Tool goes to its Home position)

T1515; (Tool no. 15 with offset no. 15 selected)

M03 S2000; (Spindle speed increased for the finishing cycle G70)

G00 X32 Z2; (Rapid positioning at the start point of the finishing cycle)

G70 P1 Q2 F30; (The finishing cycle, G70, removes the leftover material specified as finishing allowance, at a reduced feedrate for a better surface finish)

M05; (Spindle stops)

G28 U0 W0; (Tool goes to its Home position)

M30; (Reset and rewind)

THE PROGRAM FOR A CAST / FORGED WORKPIECE

The cast / forged workpiece shown in Fig. 5 is slightly over-sized, compared to the desired shape. It can be seen that it has machining allowances of 2.5 mm in the radial direction and 3 mm in the axial direction. The most efficient method for such a job is machining by G73, for which the radial and axial relief amounts would be equal to the radial and axial machining allowances, respectively. With such a choice, the first pass of G73 just follows the initial profile of the workpiece, without removing any material. The subsequent passes gradually approach the desired final shape by shifting in the manner shown in Fig. 2 (c). The number of passes is calculated on the basis of the larger of the provided machining allowances, 3 mm in this case ($R = 3 / 0.5 + 1 = 7$). Thus, G73 can be used in the following manner:

G00 X32 Z0;

G73 U2.5 W3 R7;

G73 P1 Q2 U0.1 W0.1 F60;

N1 G00 X10 Z0;

G01 Z–15;

G02 X20 Z–20 R5;

G01 Z–40;

X30 Z–50;

N2 Z–70;

Finally, since finishing allowances are specified, G70 P1 Q2 will need to be executed with a finishing tool, preferably at a higher RPM and a lower feedrate for a better surface finish.

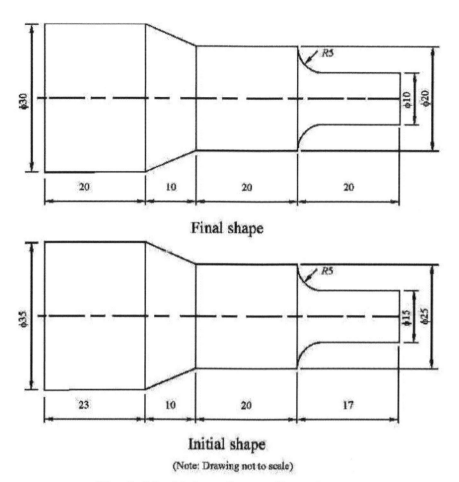

Final shape

Initial shape

(Note: Drawing not to scale)

Fig. 5: Machining of a cast/forged part

53

VOLUME 3:

Live Tool Drilling Cycles
on a
Fanuc Lathe

PREFACE

This is the third e-book in the series, "CNC Programming Skills." The first book, which was published in December 2014, dealt with the program entry and editing procedures on Fanuc machines. The second book, published in January 2015, explained the intricacies of the canned cycle G73 on a Fanuc lathe. The present book explains how to use the canned cycles for drilling/tapping/boring using live tools on a lathe with Fanuc 0i or a similar control.

A good number of jobs require machining by both lathe and milling machines. A flange with bolt holes is one example. While turning and facing are done on a lathe, the holes need to be drilled on a milling machine (or a drilling machine). Involvement of two different machines with independent setup requirement increases the cycle time as well as production cost, despite the fact that the operation on the milling machine is very simple.

To address this issue, these days higher-end lathes come equipped with live-tooling arrangement, allowing certain milling-like operations such as drilling, boring and tapping anywhere on the face of the workpiece, as well as in the radial direction at any angular position. The live tools can also be used for operations like polygon turning (e.g., to make hexagonal bolt head) and cylindrical interpolation (e.g., to make a cylindrical cam).

We will restrict our discussion to drilling/boring/tapping operations only which are very common. To facilitate these operations, a number of built-in canned cycles are available. We would be explaining these with suitable examples in quite detail so as to cover nearly everything a programmer needs to know about them.

With this book in your hand, you would not need to refer to any other book or manual for the purpose of efficiently using these canned cycles, nor would any guidance from any programming expert be needed. Read on and find out yourself. Enter the world of live tools with a bang!

LIVE TOOLS ON A LATHE

On a lathe, equipped with live tooling (which allows a tool, obviously a drilling or a similar tool, to rotate at the specified RPM, as in a milling machine) and an additional C-axis control (making it a three-axis lathe) with a provision for the C-axis clamp, canned cycles for drilling can be used for simplifying drilling, boring, and tapping (including rigid tapping) operations along X-axis as well as along/parallel to Z-axis. We will use the term "drilling" as a general term for all the three types of operations.

WORKPIECE POSITIONING AXES

The C-axis (which is called H-axis in incremental coordinate system) is a rotary axis along the Z-axis, which is controlled independently for orienting the workpiece held in the chuck at any desired angle. Thus, by manipulating C-axis and X-axis (which moves the tool to the specified radial position), a live axial drilling tool can be made to drill along Z-axis at any point on the face of the workpiece. Similarly, by manipulating C-axis and Z-axis, a live radial drilling tool can make radial holes anywhere on the cylindrical surface of the workpiece.

The main spindle has two functions: (1) continuous rotation at the specified RPM for machining such as turning, and (2) positioning itself at the specified angle for milling-like operation by a live tool. The selection between rotation mode and positioning mode (also called spindle-indexing mode) of the main spindle is done through M codes (M50/M51 are commonly used for setting spindle-indexing mode OFF/ON).

SIDE AND FRONT DRILLING

Without a Y-axis, the live drilling tools, like all other tools, can move only in XZ-plane for positioning. Thus, radial drilling is parallel to X-axis in XZ-plane, and axial drilling is parallel to Z-axis, in XZ-plane. These are referred to as side drilling and front drilling, respectively. Obviously, the axis of the tool has to be parallel to Z-axis for front drilling, and parallel to X-axis for side drilling. Table 1 lists the available canned cycles with their brief descriptions.

Table 1: Drilling canned cycles

G code	Drilling axis	Inward motion	Action at the bottom of the hole	Outward motion	Applications
G80					Cancels drilling canned cycle
G83	Z-axis	Intermittent or continuous feed	Dwell	Rapid traverse	Front drilling
G84	Z-axis	Continuous feed	Dwell, followed by spindle CCW	Continuous feed	Front tapping
G85	Z-axis	Continuous feed	Dwell	Continuous feed (2x)	Front boring
G87	X-axis	Intermittent or continuous feed	Dwell	Rapid traverse	Side drilling
G88	X-axis	Continuous feed	Dwell, followed by spindle CCW	Continuous feed	Side tapping
G89	X-axis	Continuous feed	Dwell	Continuous feed (2x)	Side boring

In front drilling (G83, G84, and G85), X and C axes are positioning axes, whereas Z and C axes are positioning axes in side drilling (G87, G88, and G89). G83 and G87, G84 and G88, and G85 and G89 have same functions – drilling, tapping, and boring, respectively – except that positioning and drilling axes are different, one suitable for front drilling and the other suitable for side drilling. And, of course, the corresponding live tool must have its axis parallel to the desired drilling axis.

LIVE-TOOL-SPINDLE ROTATION

In the absence of a live tool, there is only one spindle to rotate which is controlled by M03/M04/M05. However, if a live tool is also there, it is possible to control its spindle through these M codes also. This is machine-specific, and depends on how the machine logic (through the PMC ladder diagram) has been designed by the machine tool builder (MTB). A common method is to use the M codes for setting spindle-indexing mode OFF/ON. If the spindle-indexing mode is OFF (M50), M03/M04/M05 are for the main spindle, and if it is ON (M51), these M codes control the live-tool spindle. A less common method is to use another set of M codes for live-tool spindle, such as M13/M14 for CW/CCW rotation, as specified in parameters 5112 and 5113, respectively. When 0 is stored in these parameters, the first method is used (M03/M04 with M51). The live-tool spindle is commonly referred to as sub-spindle.

DRILLING/TAPPING/BORING CYCLES

The drilling canned cycles are modal G codes, belonging to G-code group 10, and remain in effect until canceled by G80, or replaced by some other canned cycle from the same group. In fact, once specified, all the drilling data (except the number of repeats K) are also retained as modal data, until modified or canceled by G80. Hence, only modified data need be specified in subsequent calls of these cycles, till these are canceled. Note that the drilling feedrate F is retained even after the cycles are canceled.

In general, the drilling cycles (except rigid tapping which has some difference) consists of the following operations (see Fig. 1 for a graphical representation):

Operation 1: Rapid positioning at the hole axis, i.e., at the specified (X, C) coordinate without changing Z coordinate for front drilling, or at (Z, C) coordinate without changing X coordinate for side drilling. This position is referred to as the initial level.

Operation 2: Action at initial level.

Operation 3: Rapid traverse to the R-point, moving along the hole axis, i.e., parallel to Z-axis in front drilling, and parallel to X-axis in side drilling.

Operation 4: Continuous or intermittent feed motion (hole machining) with partial or full retraction up to the R-point, till the bottom of the hole is reached.
Operation 5: Action at the bottom of the hole.
Operation 6: Feed or rapid retraction to R-point.
Operation 7: Action at R-point.
Operation 8: Rapid retraction to the initial level.

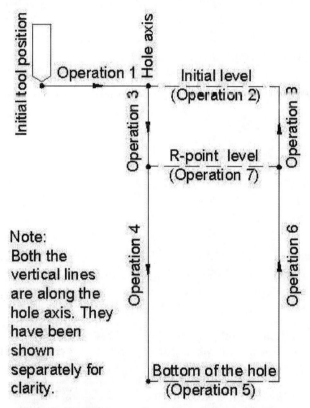

Fig. 1: Drilling cycle operation sequence

FRONT/SIDE DRILLING CYCLES (G83/G87)

On a milling machine, two separate G codes are available for full-retraction peck drilling (suitable for holes with large length to diameter ratio) and high-speed (partial retraction) peck drilling applications (G83 and G73, respectively). On a lathe, however, there is only one G code available for both types. The selection between the two types is done through a parameter. When parameter 5101#2 is set to 0, G83/G87 become high-speed peck drilling cycles. On the other hand, when this parameter is set to 1, these become full-retraction peck drilling cycles. These cycles are explained in Figs. 2 and 3, which correspond to 5101#2 set to 0 and 1, respectively. Selection of retraction-type through this parameter is valid when parameter 5161#0 is set to 0 (the default value). When set to 1, G83.5/G87.5, and G83.6/G87.6 are used for partial and full retractions, respectively. While the full-retraction cycle draws the chips out of the hole after each peck (which makes it suitable for very deep holes), the high-speed cycle clears the chips in its final retraction only, though it does break the chips and allows the coolant to flood and cool the cutting zone as well as the tool.

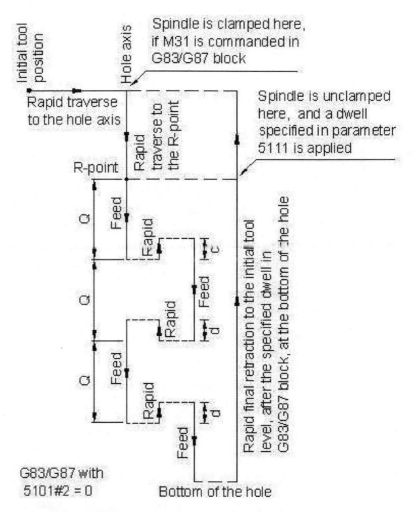

Fig. 2: Front/side high-speed peck-drilling cycles

Note the following in Fig. 2:

1. All the vertical lines are along the hole axis. They have been shown separately for clarity.

2. Peck continues till the bottom of the hole is reached. The final peck length will be less than or equal to Q, to suit the specified hole depth.

3. The retraction distance d is as specified in parameter 5114.

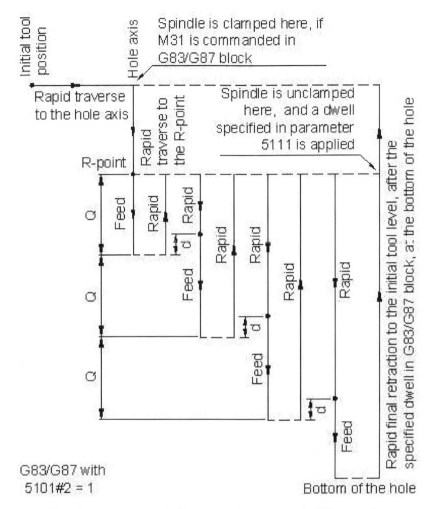

Fig. 3: Front/side full-retraction peck-drilling cycles

Note the following in Fig. 3:

1. All the vertical lines are along the hole axis. They have been shown separately for clarity.

2. Peck continues till the bottom of the hole is reached. The final peck length will be less than or equal to Q, to suit the specified hole depth.

3. All retractions are rapid.

4. Inward motion is rapid up to d distance above the previous drilled depth, followed by feed motion to increase the hole depth by Q.

65

5. The clearance distance d is as specified in parameter 5114.

Syntax

G83 X(U)_ C(H)_ Z(W)_ R_ Q_ P_ F_ K_ M_; for front drilling

G87 Z(W)_ C(H)_ X(U)_ R_ Q_ P_ F_ K_ M_; for side drilling

where

X(U)_ C(H)_ or Z(W)_ C(H)_: Hole position data
Z(W)_ or X(U): Position of the bottom of the hole
R_: Position of R-point
Q_: Peck length in micron (one-tenth of thou in inch mode)
P_: Dwell in millisecond at the bottom of the hole
F_: Feedrate
K_: Repeat count (non-modal data), if needed
M_: M code for C-axis clamp, if needed

Hole position data

The location of the axis of the hole can be specified in both absolute and incremental coordinate systems. In front drilling, (X, C) are absolute coordinates, and (U, H) are corresponding incremental coordinates. In side drilling, (Z, C) and (W, H) are, respectively, absolute and incremental coordinates. The incremental coordinates are measured from the position of the tool at the time of calling the canned cycle. In front drilling, X/U are diameter values, if diameter programming is being used. In G-code system B and C, G90/G91 with X, C, and Z are used for absolute/incremental coordinates.

Position of the bottom of the hole

Z and X are absolute coordinates of the bottom of the hole, in front drilling and side drilling, respectively. The corresponding incremental coordinates are W and U, which are measured from the R-point level, and are always negative. In side drilling, X/U are diameter values, if diameter programming is being used. Therefore, for example, if the distance between the R-point and the bottom of the hole is 10 mm, U-20 (G91 X-20 in G-code system B and C) would need to be specified, in diameter programming.

Position of R-point

In G-code system B and C, depending on certain parameter settings, R would either *always* be incremental distance from the initial level (irrespective of G90/G91), or it can be either absolute coordinate or incremental distance from the initial level (depending on G90 and G91, respectively). In system A, which we are following, this is again parameter dependent; it can be either absolute coordinate or incremental distance from the initial level. Since parameter settings are going to vary on different machines, the best way would be to execute a program on the machine, in a safe working zone, to find out whether R is absolute or incremental. Another way would be to set the parameter 5102#6 to 0, which would force R to *always* be the incremental distance from initial level, in all the three G-code systems. The incremental distance would always be negative in this case.

Another issue regarding its value, in side drilling (in front drilling, it is always the actual distance), is that whether it would be a diameter value (in diameter programming) or a radius value (even in diameter programming), depending on parameters. Therefore, either conduct an experiment on the machine to find out what it is, or set parameter 5102#7 to 0 which would *always* force it to be a radius value.

Peck length

This is specified in multiples of least input increment, without a decimal point. Thus, in millimeter mode, micron (0.001 mm) is used, and in inch mode ten-thousandth part of an inch (i.e., 0.0001 inch), is used. For example, for a peck length of 5 mm, Q5000 is programmed in millimeter mode. In inch mode, if 0.2 inch peck length is desired, Q2000 is programmed. If Q is not commanded, the entire hole is made in a single peck, converting the peck-drilling cycle into a continuous-drilling cycle.

Dwell at the bottom of the hole

If needed, e.g., for a better-machined bottom, a dwell can be specified in milliseconds without a decimal point.

Feedrate

It can be specified either in feed/minute or feed/revolution, depending on selection of feedrate mode (G98/G99, respectively, in G-code system A). The two feedrate forms are related as

Feed in mm/min = Feed in mm/rev x RPM

Repeat count

Repeating a cycle in absolute coordinate mode (X, Z, and/or C) is meaningless since the specified drilling operation would be carried out at the same place repeatedly. However, in incremental coordinate mode (U, W, and/or H), a desired number of equi-spaced holes can be very conveniently made just by a single command.

The repeat count is specified in K_, as a one-shot (non-modal) data, effective only in the block where it is commanded. Up to 9999 repeats can be specified. For a single execution, specify K1, or do not specify K at all. K0 is same as K1, if parameter 5102#5 is set to 0. When 5102#5 is set to 1, the specified modal drilling data is just stored without drilling being performed.

M codes for C-axis clamp/unclamp

After orienting the main spindle at the specified angle, it is necessary to hold it rigidly (as if in a vice) for drilling holes in the workpiece. In other words, the C-axis must be clamped. This is done through an M code, specified in parameter 5110, which applies a mechanical brake on the spindle. For example, if 31 (the usual choice) is stored in parameter 5110, M31 would clamp the spindle. The next number automatically becomes the code for spindle unclamp. Thus, in this example, M32 would release the brake. Of course, spindle unclamp at R-point, during final retraction, is a built-in feature of these cycles, obviating the need for explicitly commanding M32. In fact, this is the reason why M31 is needed in every subsequent block of these cycles (for making holes at other locations). Note that, for light machining applications, mechanical clamping of the spindle is not needed. In fact, M31 should not be commanded unless it is absolutely necessary, since it increases the cycle time.

Final retraction after hole machining

There is some difference in the way these cycles are commanded/behave in different G-code systems. The description here refers to system A. System-B and system-C cycles are similar to canned cycles on milling machines, with provision for selection between R-point retraction and initial-level retraction with G99 and G98, respectively. In system A, the final retraction is up to the initial level, if parameter 5161#1 is set to 0 (its default value). When this parameter is set to 1, the final retraction is up to the R-point. Figures 2 and 3 assume that this parameter stores 0.

Cancellation of canned cycles

Apart from the cancellation code G80, which is the usual and recommended method, these cycles can also be canceled by commanding a G code belonging to group 1 (G00, G01, G02 and G03).

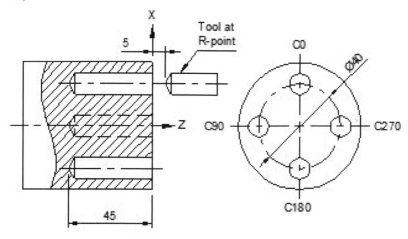

Fig 4: Offcenter front drilling on a lathe

Sample Program

Consider the job shown in Fig. 4, as an example of offcenter front drilling on a lathe. The C0 position is assumed to lie along the X-axis. Thus, the positions of the four holes are (X40, C0), (X40, C90), (X40, C180), and (X40, C270), respectively. The incremental coordinates would be (U-10, H0), (U0, H90), (U0, H90), and (U0, H90), respectively, if the initial tool position is (X50, C0). U0 need not be explicitly commanded. The program for this job can be written as

M51; (C-axis indexing mode ON)
G00 X50 C0 Z20; (Initial position of the tool)
M03 S1000; (Live spindle starts)
G83 X40 C0 Z–45 R–15 Q5000 P1000 F10 M31;
(First hole drilled; C0 not needed)
(or G83 U-10 H0 W-50 R-15 ...)
C90 M31; (Second hole drilled)
(or H90 M31)
C180 M31; (Third hole drilled)
(or H90 M31)
C270 M31; (Fourth hole drilled)
(or H90 M31)
(In fact, the last three holes can be drilled by a single command H90 K3 M31)
G80; (Canned cycle canceled)
M05; (Live spindle stops)
M50; (C-axis indexing mode OFF)

It is also possible to drill all the four holes by a single G83 command:
G83 X40 H90 Z–45 (or W–50) R–15 Q5000 P1000 K4 F10 M31;
which would drill the second hole first, followed by third, fourth and first holes (with CCW positioning of the live tool with respect to the face of the workpiece. The workpiece actually rotates clockwise for the required positioning while the live tool remains stationary). And, without a Q-word, all the holes would be drilled with continuous feed, i.e., without pecks.

FRONT/SIDE TAPPING CYCLES (G84/G88)

These cycles are similar to G83/G87 (without a Q-word), when run in non-rigid mode which is also referred to as standard or conventional mode (see Fig. 5). The explanation of all the words is same. The feedrate in mm/min is calculated using the formula

Feed in mm/min = Pitch in mm x RPM

which is used in G98 (feed per minute) mode. In G99 (feed per revolution) mode, the pitch of the tap is specified as feedrate.

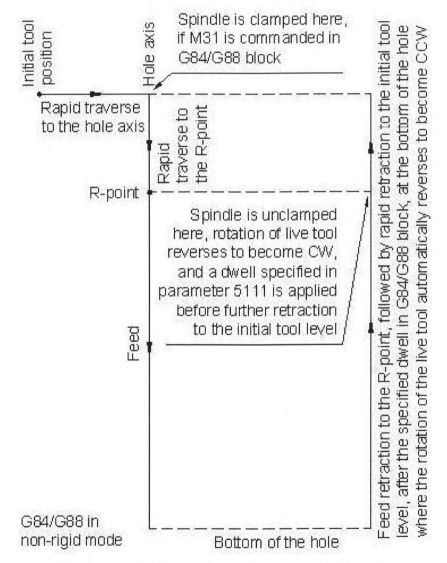

Fig. 5: Front/side tapping cycles in non-rigid mode

Note the following in Fig. 5:

1. Both the vertical lines are along the hole axis. They have been shown separately for clarity.

2. G84/G88 can be run both in normal mode and rigid mode. Here, normal mode is assumed. In normal mode, a floating tap holder must be used.

3. M03 must be commanded before invoking G84/G88. Left-hand tapping is not available.

4. Feedrate override switch remains disabled during tapping.

Syntax

G84 X(U)_ C(H)_ Z(W)_ R_ P_ F_ K_ M_; for front tapping

G88 Z(W)_ C(H)_ X(U)_ R_ P_ F_ K_ M_; for side tapping

where

X(U)_ C(H)_ or Z(W)_ C(H)_: Hole position data
Z(W)_ or X(U): Position of the bottom of the hole
R_: Position of R-point
P_: Dwell in millisecond at the bottom of the hole
F_: Feedrate
K_: Repeat count (non-modal data), if needed
M_: M code for C-axis clamp, if needed

Operation in rigid mode

These cycles can also be run in rigid mode (see Fig. 6), obviating the need for using a floating tap holder, and with the added advantage of reduced cycle time, since higher RPM as well as up to twenty times higher retraction speed can be programmed in rigid mode. Rigid mode can be selected either by commanding M29 (which is the number stored in parameter 5210 in the range 0 to 255. If 0 or 29 is stored, M29 selects the rigid mode) in or before the tapping block, or by setting parameter 5200#0 to 1, which would always use rigid mode for G84/G88, as on a milling machine. Note that when a rigid cycle ends, the live-tool spindle remains stopped, as if S0 is commanded. Therefore, an S-value would need to be re-specified if spindle rotation is again needed.

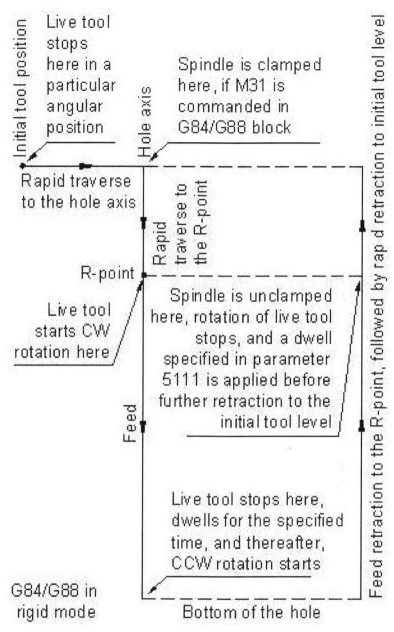

Fig. 6: Front/side tapping cycles in rigid mode

Note the following in Fig. 6:
1. Both the vertical lines are along the hole axis. They have been shown separately for clarity.

73

2. In rigid mode, floating tap holder is not needed. Rigid taps are used.

3. M03 need not be commanded before invoking G84/G88, but an S-word is needed. Spindle-stop and CW/CCW-start at the specified RPM, as and when needed, is a built-in feature of the cycle.

4. Left-hand tapping is not available.

5. Both feedrate override and spindle override switches remain disabled during tapping.

Retraction speed

The default speed is the specified feedrate. Override, in terms of percent of specified feedrate, is permitted if parameter 5200#4 is set to 1. With this setting, the override percentage is determined by parameters 5201#3 (valid data is 0 or 1) and 5211 (valid data is an integer in the range 1 to 200). The value stored in 5211 is used as percent override if 5201#3 is set to 0, giving 1%, 2%, ... 200% override percentages. When 5201#3 is set to 1, these get multiplied by 10, giving 10%, 20%, ... 2000% overrides. Thus, the retraction speed can be chosen to be up to 20 times (corresponding to 2000%) the specified feedrate.

Sample program

Consider the same job of Fig. 4, with pre-existing holes of required diameter. Assuming 1.5 mm pitch and 1000 RPM, the required feedrate would be 1500 mm/min or 1.5 mm/rev, depending on the feedrate mode. For feed per minute mode, the program can be written as

<Standard-tapping program; requires floating tap holder>
G98; (Feed per minute mode)
M51; (C-axis indexing mode ON)
G00 X50 C0 Z20; (Initial position of the tool)
M03 S1000; (Live spindle starts)
G84 X40 H90 Z–45 (or W–50) R–15 P1000 K4 F1500 M31; (Tapping done)
G80; (Canned cycle canceled)
M05; (Live spindle stops)
M50; (C-axis indexing mode OFF)

<Rigid-tapping program; does not require floating tap holder>
G98; (Feed per minute mode)
M51; (C-axis indexing mode ON)
G00 X50 C0 Z20; (Initial position of the tool)
M29 S1000; (Rigid-tapping mode selected)
G84 X40 H90 Z–45 (or W–50) R–15 P1000 K4 F1500 M31;
(Rigid tapping done)
G80; (Canned cycle canceled)
M50; (C-axis indexing mode OFF)

Peck tapping

Continuous tapping operation up to the bottom of the hole does not clear the chips until final retraction. While this creates no problem for shallow holes, the tapped hole and/or the tap itself may get damaged due to chip clogging, if the hole is too deep (actually, when the depth to diameter ratio is large). In such cases, peck tapping solves the problem, which works in a way similar to hand tapping – we tap up to some distance, then retract a bit by rotating the tap in the opposite direction, then tap further, retract again, and continue in the same manner till the bottom of the hole is reached. A similar thing can be done with G84/G88 also, in both rigid and non-rigid modes, by specifying a Q-word for peck length, as in a peck drilling cycle, provided parameter 5104#6 is set to 1. When this parameter is set to 0, a Q-word in G84/G88 becomes invalid. When set to 1, and pecking is not desired, do not specify a Q-word or specify Q0.

As in case of peck drilling, two types of retractions after every peck is possible: partial retraction (high-speed pecking) and full retraction up to the R-point (standard pecking, suitable for deep holes), selectable through parameter 5200#5 when set to 0 and 1, respectively. The retraction/clearance distance d (as defined in Figs. 2 and 3) during pecks is specified in parameter 5213.

FRONT/SIDE BORING CYCLES (G85/G89)

These cycles are similar to continuous drilling cycles (G83/G87 when used without a Q-word). The syntax as well as function/interpretation of all the words is exactly same. The only difference is in the nature of final retraction. In G83/G87, it is rapid, whereas it is at a controlled rate in G85/G89. The retraction speed in boring cycles is parameter dependent:

5104#1 = 0 causes retraction at a controlled rate.
5104#1 = 1 causes retraction at rapid traverse rate.

When parameter 5104#1 is set to 1, G85/G89 become same as G83/G87 (without a Q-word), making it meaningless. Therefore, this parameter must be set to 0 which is also its default value. When set to 0, the speed of retraction is determined by parameter 5121. The retraction speed becomes equal to the specified feedrate multiplied by the value stored in this parameter, the valid data range being 0.1 to 20. Thus, the retraction speed can be chosen to lie between 0.1F and 20F where F is the specified feedrate. For a value higher than 20, 20 is assumed and 20F is used. And, a 0 value results in 2F speed. Since the default values of both the parameters are 0 only, and people generally do not care to change these parameters, the speed of final retraction in a boring cycle is most likely to be twice the feedrate.

VOLUME 4:

Understanding Offsets
on
Fanuc Machines

PREFACE

This is the fourth e-book in the series, "CNC Programming Skills." It explains the concept of offsets and how to measure these manually on i-series Fanuc machines or a similar control.

The very first thing which a person, who is new to the CNC world, needs to know is how to setup the coordinate system for machining. The machining coordinate system is defined by adding relevant offsets to the pre-defined machine coordinate system. When it comes to offsets, oh! these are just too many to confuse most people: External offset, Work offset, Geometry offset, Wear offset, Length offset, to name a few. And, these are to be understood in totality, as these are interrelated. Certain things are illogical also such as the wear value of the tool tip number on a lathe. Unfortunately, we cannot question Fanuc. We have to understand their language if we want to survive in the CNC world.

This book is written keeping in view the requirements and expectations of beginners who wish to learn on their own with no helping hand available to them. No background is assumed. One who has at least seen the operator's panel and the manual data input panel, and knows elementary things like the directions of axes on a CNC machine, would be able to follow the book. Starting from the very basic, it has everything a seasoned CNC operator is supposed to know. Read on and find out yourself. Enter the world of CNC with confidence!

HOW TO USE THE BOOK

Since the offsets are generally interrelated, the concept has to be understood in totality. Skipping the initial pages and jumping to a later topic may result in some missing links, leading to lack of clarity. It is, therefore, suggested to read the book from the very beginning without skipping anything. If something is not clear in the first reading, just ignore it for the time being and proceed further. Come back to it after you have finished the book once. It is not a difficult topic. The main problem is that things are generally not explained methodically elsewhere. As a result, almost every CNC operator handles offsets in his own way. Even though the final result might be correct, the adopted methods do confuse a new learner as to which method is the proper one. The method presented in the book is based on what most people are doing. The author believes that one weekend is sufficient to understand the concepts.

WHAT EXACTLY AN OFFSET IS

The machine uses a pre-defined coordinate system which is called the *machine coordinate system* (MCS). The programmer, however, writes a program in a *chosen* coordinate system which is called the *workpiece coordinate system* (WCS). For example, the origin of the WCS on a lathe is usually placed at the center of the face of the workpiece. (The programmer's coordinate system is actually the WCS plus the required tool offset added to it, which we may call a compensated WCS, as we will see later. Tool offset is explained later. We are, at this stage, considering a case where tool offset is assumed to be zero. Complexities would be introduced gradually which would keep the learning easier and smooth).

So, for the purpose of a simplified discussion, let us first consider the case of machining with just one tool with zero tool offset. When the offset or the deviation between the two coordinate systems (the displacement vector *from* the origin of the MCS *to* the origin of the WCS), which is called *work offset* vector, is entered in the memory of the machine, the machine defines such a WCS (by placing its origin at the tip of the offset vector, with respect to the MCS), and when this WCS is made active (several WCSs can be defined on a machine out of which one is selected to be active at any time) in the beginning of a program, the machine starts interpreting the coordinate values specified in the program in this WCS, as desired. (We have deliberately not talked about the external coordinate system (ECS) at this stage to keep the discussion simple. Just assume that there is no such coordinate system, for the time being).

The work offset values for different axes are the components of the work offset vector along the corresponding axes, which are fed to the machine by editing the work offset table on the "work" screen of the machine. The work offset table (in fact, all offset tables) on a 2-axis lathe has X and Z components, whereas that on a 3-axis milling machine has X, Y and Z components.

Things are not so *straightforward*, however. The complication arises because every tool, in general, has a different shape as well as size. Let us first consider the example of a lathe.

When the current tool is changed by another tool, with the turret at the *same* position, the reference points of the two tools do *not* come to the same point (The reference point of a tool is the point which is used for programming the movement of the tool. This is explained in more detail later). This means that when the reference points of the different tools are brought to the same location in space, the coordinates are all different, whereas these should have been same. This necessitates suitable adjustment in the WCS for each tool, to take care of such differences. This is done by adding what is called *geometry offset* (it gets its name because of the fact that it takes care of differences in geometry) to the work offset, algebraically (the offset values can be negative also). The geometry offset table appears on the "geometry" screen of a lathe.

The story is not over yet. Even with the same tool, minor differences in machining could be observed because of various reasons such as a changed machining condition or tool wear in the course of time. This is taken care by further adding what is called *wear offset.* Geometry offset is used for taking care of major deviation, whereas wear offset is for minor deviation. These two have been provided separately for convenience. Both have both X component (value in the radial direction) as well as Z component (value in the axial direction). The wear offset table appears on the "wear" screen of a lathe. The work screen, the geometry screen and the wear screen are selectable through the corresponding soft keys (WORK, GEOM and WEAR, respectively).

By the term tool offset, we mean geometry offset and wear offset on a lathe. The net offset used by the machine for establishing the compensated WCS for a particular tool, is the algebraic sum of external offset, work offset, geometry offset and wear offset. As mentioned earlier, by compensated WCS we mean the original WCS defined by external offset and work offset, with geometry and wear offsets added to it. The coordinate system thus obtained is used by the machine for machining with the selected tool. We actually need to select a tool as well as appropriate geometry and wear offsets for machining.

Essentially, the WCS establishes a baseline coordinate system, and geometry and wear offsets are added to it to establish the coordinate system to suit a particular tool. There is no specific name assigned to this coordinate system. We are calling it the compensated WCS for a particular tool.

In addition to these, there is a *local offset* also, corresponding to a *local coordinate system* which can be optionally defined inside a program, with respect to the WCS. It is also referred to as a *child coordinate system*, which remains effective in the program which defines it, till it is canceled (It must be canceled at the end of the program). It is useful when the *same* tools are to be used for machining workpieces of different sizes in the *same* WCS (Each WCS is suitable for a workpiece of a specific size. When the number of workpiece-sizes is more than the number of available WCSs on a machine, the use of local coordinate systems is a simple solution).

The deviation in the size of a different workpiece with respect to the original workpiece (with reference to which the compensated WCS was established) is specified as the local offset, to define a local coordinate system for the second workpiece. In fact, several local coordinate systems can be defined, one-by-one, (i.e., defined and redefined) to suit workpieces of different sizes in the same WCS.

The relationship between various offsets is shown in Fig. 1, in which only three out of the standard six WCSs have been shown, and geometry offset and wear offset are not shown. Conceptually, what is shown, establishes baseline coordinate systems; geometry and wear offsets are additionally added to take care of deviation in tools / workpiece sizes. In fact, even the same tool can be used to machine workpieces of different sizes, by selecting different tool offsets.

Figure 1 is valid for coordinate systems on a milling machine also where, as we will see later, tool length compensation is used to take care of variation in tool length and workpiece sizes.

Apart from these offsets, one also needs to specify the radius of the tool tip as well as its orientation with respect to the chuck, which the machine needs to know for machining with radius compensation. To have consistency with the earlier approach, the radius is specified with both geometry component (containing the actual value) as well as the wear component (containing the deviation from the actual value). In fact, even though meaningless, even the direction of the tool tip has both geometry component as well as wear component, just for the sake of a consistent approach. Actually, both refer to the same memory register, Therefore, the geometry value automatically overwrites the wear value, and vice versa.

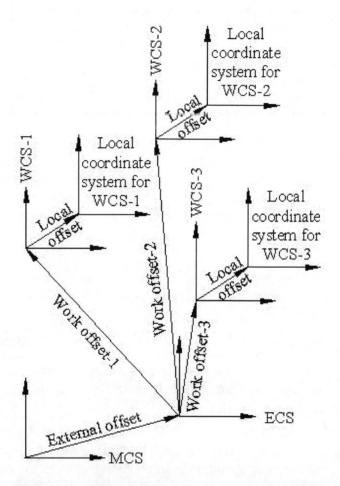

Fig. 1: Coordinate systems and offsets

On a milling machine, there is some simplicity. All tools have cylindrical geometry, differing only in length and diameter, and all are clamped in the spindle in the same manner. Therefore, when a tool is changed, XY coordinates remain the same. This requires adjustment for length and diameter only, for different tools or workpiece sizes.

On a milling machine, the term tool length compensation is often used instead of tool length offset. The terms offset and compensation are sometimes used interchangeably, with the same meaning. Thus, on a milling machine, the Z0 position, for the purpose of interpreting the coordinates used in the program, is determined by adding external offset, work offset and tool length offset (compensation). XY datum is determined *solely* by external offset and work offset for all the tools.

Several WCSs are available on a milling machine (also on a lathe and other CNC machines) to set the XYZ (assuming a 3-axis machine) datum at different places, to suit the workpieces clamped at different places on the machine table. Both length and diameter offsets have geometry component (containing the actual values) as well as wear component (containing minor deviation from the actual values). The geometry and the wear components of the length offset and the diameter offset appear on a combined offset screen on a milling machine. We will now discuss each type of offset in greater detail.

EXTERNAL OFFSET

It is possible to define an *external coordinate system* between the machine coordinate system and the workpiece coordinate systems, such that the workpiece coordinate systems are defined with respect to the external coordinate system, and the external coordinate system is defined with respect to the machine coordinate system, as shown in Fig. 1. The displacement vector *from* the origin of the machine coordinate system *to* the origin of the external coordinate system is called *external offset*. Only one external coordinate system can be defined for all the workpiece coordinate systems available on a machine. It is possible to define up to six (more than six is an optional feature of the control) different workpiece coordinate systems to suit six workpieces of different geometries / sizes. Out of these, one is chosen to be the active WCS at a time which the machine uses for coordinate calculations.

The external coordinate system is generally not used – its offset values (components of external offset vector along different axes) are kept zero, which has the effect of coinciding the external coordinate system with the machine coordinate system. The external coordinate system is, however, very useful when the chuck is replaced by a different chuck which holds the same workpiece in a different position, compared to the holding position of the previous chuck. In case of a milling machine, this situation may arise when the same fixture is placed somewhere else on the machine bed, with respect to its original position. In such a case, the difference in the new and the old holding positions of a given workpiece is specified as external offset (the displacement vector *from* the old position *to* the new position). This corrects the offset values for *all* the tools for *all* the workpieces; there is no need to repeat the offset setting process for individual tools with respect to individual workpieces. Because of this reason, the external offset is sometimes referred to as *chuck offset* on a lathe, and *fixture offset* on a milling machine.

WORK OFFSET

Work offset is the displacement vector *from* the origin of the external coordinate system *to* the origin of the corresponding workpiece coordinate system. For every workpiece coordinate system, an appropriate work offset is required to be specified which, in turn, defines such a coordinate system.

If the external coordinate system is not used, i.e., if the external offset values are kept zero, then the work offset becomes the displacement between the origins of the machine coordinate system and the corresponding workpiece coordinate system.

The appropriate workpiece coordinate system can be selected by G54 through G59. The default is G54, but it is a good practice not to rely on defaults, because it is possible to change certain default settings of the machine through changes in certain system parameters. Therefore, in general, whatever is required must be commanded explicitly in the program.

For determining the offset distances for, say, G58, G58 must be executed in the MDI mode in the beginning of the offset procedure (to make it the active WCS), and if a coordinate value used in the program is required to be interpreted in G58, G58 must be explicitly commanded in a previous block. This G code is a modal code, hence remains effective until changed by some other code from the same group (G54 through G59).

In the discussion that follows, it has been assumed that the external offset is zero.

Both lathe and milling machines show external and work offsets in a similar manner, on two screens (because of space limitation). These typically appear as shown in Fig. 2 (shown for a lathe).

In both the screens, G54 at the top left corner indicates that the G54 work coordinate system is active. For measuring the work offset for this coordinate system, the curser should be brought to highlight the X or Z display (anyone) of G54 before measuring (through the MEASUR soft key) the offset distances. Offset distances for the other coordinate systems (G55 through G59) *cannot* be measured in the G54 mode. This applies to all the workpiece coordinate system; offset distances for a particular WCS can be measured *only when* it is the active WCS.

For measuring the offset distances for a different coordinate system (say, G58), the curser should be brought to highlight the X or Z display of G58 *after* executing G58 in the MDI mode. The display at the top left corner will change to G58 to indicate the G58 mode.

Data for the external coordinate system can be provided in any WCS mode. However, as already discussed, these are generally kept zero.

The work offset screen and the offset measurement method on a milling machine are exactly same as described above, except that there are extra rows for the additional axes.

WORK COORDINATES
(G54)

NO.		DATA		NO.		DATA
00	X	0.000		02	X	
(EXT)	Z	0.000		(G55)	Z	

NO.		DATA		NO.		DATA
01	X			03	X	
(G54)	Z			(G56)	Z	

First screen

WORK COORDINATES
(G54)

NO.		DATA		NO.		DATA
04	X			06	X	
(G57)	Z			(G59)	Z	

NO.		DATA
05	X	
(G58)	Z	

Second screen

Fig. 2: Work offset screens on a lathe

THE CONCEPT OF MASTER TOOL

It is convenient if the work offset on a lathe is measured with respect to a turning tool which can, therefore, be called the *master tool*. For such a choice, there would be no deviation between the origin of the WCS and the reference point of this tool when it is brought to the (0, 0) position (as displayed in this WCS). Consequently, the geometry offset of the master tool becomes zero. Wear offset, of course, needs to be used to compensate for minor variations in machining.

TOOL CALL ON A LATHE

The T word on a lathe is a four digit word which causes tool change with the desired tool offsets. Usually, depending on a system parameter, the two digits towards the right indicate both the geometry offset number and the wear offset number, and the remaining digits towards the left specify the tool number. Thus, for example, if the master tool is tool number 1, T0101 (or T101) calls the master tool with geometry offset number 1 (which stores zero values for X and Z offsets) and wear offset number 1. (An offset number refers to the row number of the respective offset table on the offset screen.)

OFFSET MEMORY TYPES ON A LATHE

For a lathe, *Memory Type A* and *Memory Type B* are available for storing tool offsets. Type B is more advanced, and used on recent controls. Our discussion on lathe offsets pertains to Type B memory.

GEOMETRY OFFSET ON A LATHE

When another tool (say, tool number 2) is required to be used, and the turret places it in the cutting position, its tip will not come exactly at the position of the tip of tool number 1, the master tool, because of a possible difference in the geometries of the two tools. This difference in the tip positions (the components of the vector *from* the master tool *to* tool number 2), which can be either positive or negative, are specified as X and Z geometry offsets for tool number 2. This needs to be entered in the geometry offset table in a chosen row. For convenience in referencing, we use row number 2 (which is called offset number 2) of the geometry offset table for entering the offset values for tool number 2. Then, the program will call tool number 2 by T0202. Geometry offset values need to be ascertained for all the tools which are to be used. The program calls the tools by T0303, T0404 etc.

The number of rows in the offset table (i.e., offset numbers) is generally a lot more than the number of available tools on a machine. This is provided to allow a tool to be used with more than one offset, to suit workpieces of different lengths.

WEAR OFFSET ON A LATHE

With time, the tip of a tool starts wearing out. As a result, external diameters increase in size and internal diameters become smaller. Facing also becomes slightly oversize. In fact, the entire machining operation gets affected. (A changed machining condition also may result in slight variation in dimensions.)

Though this error can be corrected by modifying the geometry offset value of that tool, it is not considered a good practice because we will lose the original value, and, as a result, when it is decided to change the insert, the geometry offset setting will have to be done again, even though the new insert may be exactly same as the previous one. Therefore, any inaccuracy introduced on account of tool wear (or any other reason such as a changed machining condition) is specified as wear offset. When the insert is replaced by a new one, the *same* geometry offset is used, and the wear offset is reset to zero. Minor inaccuracy, if any, because of any reason, should be compensated by specifying appropriate wear offsets (even though the insert is new!). Actually, the purpose of wear offset is to take care of minor inaccuracies caused by various reasons including tool wear.

OFFSET NUMBER ON A LATHE

Usually the last (rightmost) two digits of a T word specify the geometry as well as wear offset numbers (the row numbers of the respective offset tables), and the remaining digits (one or two) at the left designate the tool number on a lathe. Though this is what we are going to assume in further discussion, the numbering scheme for geometry offset, however, depends on a parameter setting. Though the leftmost two digits always designate the tool number, a different parameter setting will cause these digits to be interpreted as geometry offset number also. The rightmost two digits are always wear offset numbers. So, the relevant parameter setting must be verified.

Alternatively, conduct a small experiment: execute, say, T0203 in the MDI mode and verify whether the machine uses 02 or 03 as the geometry offset number, by bringing the tool to a known position and comparing it with the coordinate display on the screen of the machine. If you measured X and Z geometry offsets for this tool while the row number 03 of the geometry offset table was highlighted, and the current position displayed on the screen is correct, then the machine is using the rightmost two digits as geometry offset number. The best thing, however, would be to use the same numbers for both the leftmost and the rightmost pair of digits, such as T0202. This will always have the same meaning, irrespective of the parameter setting. It is also possible to use only two digits with a T word on a lathe. This again depends on a parameter setting. This method, however, is not commonly used.

In the discussion that follows we would be assuming that the row number for both geometry offset and wear offset is same. For example, T0222 (a row number different from 02 has been deliberately used for sake of illustration) uses row number 22 (last two digits of 0222) of geometry offset table as well as of wear offset table, for tool number 2.

TOOL TIP NUMBERS FOR RADIUS COMPENSATION

The basic theory of radius compensation requires that the program be written for the *center of the arc or circle*, the periphery of which cuts the material. Then, the desired left or right compensation is incorporated by G41 or G42. Therefore, if the nose center of a turning tool is made the reference point on a lathe, there is no need for the concept of tip numbers (which are also called nose numbers). In fact, in such a case, the assigned tip number is 0 (number 9 is also used).

However, the usual method of the offset setting on a lathe selects a different point as the reference point of the tool, which is called the *imaginary tool tip*, as shown in Fig. 3 (where the size to the tool tip is exaggerated for clarity). There is a difficulty that the tool tip cannot be accurately placed at the X0 position as there is no physical surface available for reference at X0. We, therefore, need to touch the surface of a workpiece of known diameter, and use the diameter as the X value, for indirectly locating the X0 level. (We are assuming that the diameter programming is being used on the lathe, rather than radius programming, In diameter programming, the X coordinate of a point equals twice the radial distance from the axis.) For the Z reference, the tool is made to touch the face of the workpiece which is the Z0 position.

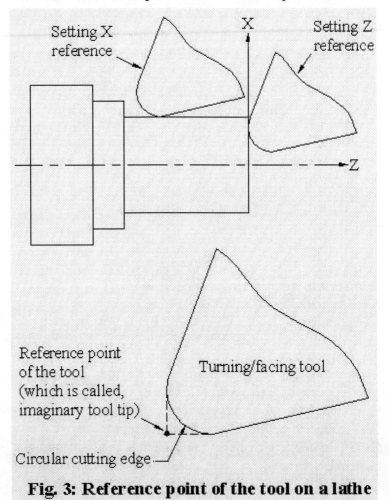

Fig. 3: Reference point of the tool on a lathe

The implication of this method of offset setting is that the reference point of the tool (imaginary tool tip) does not lie on the body of the tool. Even though this point controls the movement of the tool, it never cuts any material. The bottom of the tool cuts in straight turning, the side of the tool cuts in straight facing, and some point (which keeps shifting) on the circular cutting edge cuts in circular interpolation or taper motion.

It should now be clear why, even without using radius compensation, there is no error in straight turning or straight facing if this point is used as the reference point, which is a great advantage (Slight error at the *end* of the cutting stroke would be there which would usually be insignificant). The error in circular or taper machining due to the circular cutting edge can be corrected by invoking radius compensation. On the other hand, if the nose center is made the reference, one cannot machine without invoking radius compensation because it would result in overcutting even in straight turning or straight facing, which is a serious drawback.

The control, however, needs to know where the center of the arc lies, for incorporating radius compensation. It can be seen in the Fig. 4 that, for any type (left hand, right hand or neutral) and any orientation (external or internal, face up or inverted position) of the tool, the magnitude of the distances along X and Z axes between the reference point and the center are always equal to the radius of the arc (or zero, in some cases). Therefore, in order to completely locate the center, it is sufficient to know the direction (orientation) of the center with respect to the reference point of the tool. The calculation for locating the center point is done by the control, on the basis of the specified nose radius and its direction. Every possible direction has a tip number (also called, nose number) assigned to it for identification, as shown in the figure. The tip number is entered in the fourth column (which has the heading T) of the geometry or wear offset table. The program is always written for the usual reference point only, *with or without* radius compensation.

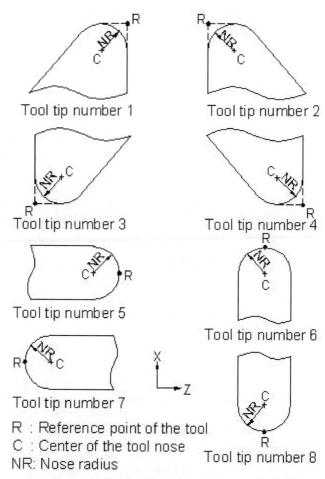

Fig. 4: Tool tip numbers on a lathe

Consider, as an example, turning from right to left using a right hand tool, kept in the inverted position on a rear type lathe with clockwise spindle (M03). From radius compensation point of view, this case is equivalent to turning from right to left using a left hand tool in the normal (i.e., non-inverted) position on a rear type lathe with counter clockwise spindle (M04). The machine recognizes such an orientation of the tool by tip number 3 (refer to Fig. 4), which should be specified in the offset table. In this orientation, the center is located towards right and upwards with respect to the reference point of the tool. So, the Z coordinate of the center can be calculated by *adding* the nose radius to the Z coordinate of the reference point. Similarly, the X coordinate of the center would be equal to the X coordinate of the reference point *plus* twice the nose radius. The calculation is done by the machine automatically in the radius compensation mode.

As another example, consider turning (towards left or right) by a neutral type tool on a rear type lathe. The machine recognizes this case by nose number is 8, and calculates the X coordinate of the center by *adding* twice the nose radius to the X coordinate of the reference point, while using the *same* Z coordinate.

Thus, it is not necessary to change the usual method of offset setting (as shown in Fig. 3) for selecting the center of the nose as the reference point of the tool, for the sole purpose of incorporating radius compensation. In fact, this is not desirable because of reasons already explained. The machine is simply given the information about the type / orientation of the tool through the appropriate nose number. This allows the machine to automatically locate the center of the tool tip, for the purpose of radius compensation.

GEOMETRY AND WEAR OFFSET SCREENS ON A LATHE

The geometry and the wear offset screens typically appear on a lathe as shown in Fig. 5.

Typically, 64 rows (hence 64 offset numbers) are available in these offset tables, with an option to increase the number (on extra payment, of course). In the offset numbers (row numbers), the letters G and W denote geometry and wear, respectively, just to remind the user of the active offset screen (geometry or wear). X, Z, R and T designate X axis offset (geometry / wear), Z axis offset (geometry / wear), tool nose radius (geometry / wear) and imaginary tool tip numbers, respectively. Also note that since both the offset screens must have the same tool tip number, in the last column of a particular row, change in one screen automatically results in a change in the other screen. So, the one specified last becomes effective.

OFFSET / GEOMETRY				
NO.	X	Z	R	T
G01				
G02				
G03				
...				
...				
...				
G64				

Geometry offset screen

OFFSET / WEAR				
NO.	X	Z	R	T
W01				
W02				
W03				
...				
...				
...				
W64				

Wear offset screen

Fig. 5: Geometry / wear offset screens on lathe

OFFSET MEMORY TYPES ON A MILLING MACHINE

Fanuc controls have introduced three, progressively more advanced, types of memory to store tool length and tool radius offset values on a milling machine. These are known as *Memory Type A*, *Memory Type B* and *Memory Type C*. Type C, which is the latest and the most convenient to use, is described below.

TOOL OFFSETS ON A MILLING MACHINE

A simplicity on a milling machine is that all tools have cylindrical geometry. Different tools differ only in the length and the cutter diameter. Therefore, apart from external offset and work offset, only four offset entries are needed for any tool – geometry and wear offset values of both length and radius, which are called tool offsets.

The geometry offset of length is the *difference* between the length of the actual tool and that of the *master tool* (the tool with which the work offsets are measured), whereas the geometry offset of radius is the effective cutting radius of the tool. The wear offset values are used to nullify the minor errors introduced due to a changed machining condition or due to changes in the length and / or radius of the tool because of tool wear.

OFFSET SCREEN ON A MILLING MACHINE

Since the screen size is rather small, only four columns can be conveniently displayed on it. Therefore, whereas two different offset screens (geometry screen and wear screen) are used to specify all the eight tool offset values on a lathe – four in each screen corresponding to X, Z, R and T values – only one screen is needed to specify all the four tool offset values on a milling machine. The external offset and work offsets appear in two separate screens on both the machines. A combined geometry and wear tool offset screen on a milling machine appears as shown in Fig. 6.

Note that, in the D column, the *radius* of the tool is specified, not the diameter (*radius* in geometry column and *change in radius* in wear column is specified). If radius compensation is not required to be used, only the H values need to be specified.

OFFSET	H		D	
NUMBER	GEOMETRY	WEAR	GEOMETRY	WEAR
01				
02				
03				
...				
...				
...				
64				

Fig. 6: Tool offset screen on a milling machine

TOOL CHANGE ON A MILLING MACHINE

The method of calling a tool and using its offset values is different on a milling machine. The command T01 M06 will simply place tool number 1 in the machine spindle (the T word on a milling machine is a two digit code), without recalling its length and radius offset values (Some milling machines require that T01 be commanded in a previous block, before the M06 command). These are incorporated by an H word (e.g., H01) and a D word (e.g., D01), respectively. These words need to be coupled with a tool length compensation codes, G43 or G44, and a radius compensation code, G41 or G42, respectively, to take effect.

TOOL LENGTH COMPENSATION ON A MILLING MACHINE

If the length of tool number 2 (say) is smaller than that of the master tool (say, tool number 1), then this difference can be specified as a positive or a negative value in the second row (preferably the same row as the tool number should be selected, but any other row also can be used) of the first column of length offset (H). If a negative value is specified, the length compensation is incorporated by G43. If a positive value is specified, the length compensation is done by G44. Assuming a negative specified value in the second row in this case, tool number 2 can be accurately placed at a desired height of, say, 2 mm by the following codes:

G43 H02;
G00 Z2;

These two lines can be combined into a single line also, because the two G codes belong to different groups. Further, if G00 is already active, we need to simply write

G43 H02 Z2;

in which the Z word can also precede the H word, because the order of the different words is usually not important in the *word address format* which the CNC machines use.

RADIUS COMPENSATION ON A MILLING MACHINE

If machining with radius compensation is desired, so as to cut exactly along the defined profile, by shifting the toolpath towards left or towards right with respect to the defined profile, the radius of the tool will have to be specified in the second row (say) of the first column of the radius offset (D). Then, the radius compensation (i.e., lateral shift in the toolpath) towards left (while viewing in the direction of the toolpath) is incorporated by G41, and towards right by G42.

Since the machine calculates the shift amount and its direction mathematically, a negative value for the radius in the offset table will reverse the shift directions for G41 and G42. Assuming that a positive radius value is entered in the second row, compensation towards left in the XY plane (Compensation is possible in YZ and ZX planes also, which are selected by G19 and G18, respectively. G17 is for XY plane and is the machine default.) can be incorporated by the following codes:

G41 D02;
G01 X_ Y_ F_;

where it is possible to combine these two lines into a single line. And, if G01 is already active, one can simply write

G41 D02 X_ Y_ F_;

This is called the start-up or lead-in motion during which the machine gradually incorporates radius compensation in a ramp manner, such that after this motion is complete, the tool positions itself correctly at the start point of the next move command so as to subsequently cut correctly with radius compensation. The compensation is properly incorporated only at the end of this motion. Therefore, actual cutting should not be done in the start-up move. Also, some controls require that the start-up move be larger than the radius of the tool. G00 also can be used in the start-up move, but G02 or G03 cannot be used.

Note that, for the purpose of radius compensation, the left or the right direction is defined when we look at the toolpath from the positive side of the third axis. This does not cause any confusion while machining in the XY plane because we indeed look at this plane from the positive side of the Z axis. For machining on other planes, one needs to be careful. For example, one has to look at the YZ plane from the positive side of the X axis, i.e., from right to left on a VMC. For XZ plane, one needs to look from *behind* the machine which is opposite to our normal viewing direction. Radius compensation is not possible on any other plane, though G00 and G01 both allow 3-dimensional motion.

Note that when radius compensation is required to be used on a lathe, no D word or its equivalent is used with G41 or G42. The complete information about the tool geometry, including its radius, is automatically extracted by the machine when the tool is called with an offset number. The method of incorporating tool offsets on a milling machine is different and a bit more complex, compared to that on a lathe. The requirement of the startup move after invoking radius compensation is exactly same on both the machines.

ENTERING OFFSET VALUES INTO OFFSET TABLES

The procedure on a lathe as well as on a milling machine is similar. The basic procedure involves bringing the tool to the origin (or to a known coordinate position) of the desired coordinate system, and then use the measure feature of the control (through the MEASUR soft key) to enter the offset values automatically. Let us first take the example of determining work offsets on a lathe:

• Measure the diameter of the workpiece for which the work offset is to be determined, and place it in the chuck at the desired axial position.

• Select the MDI mode, and execute the desired work coordinate system code (G54 through G59) for which work offset is to be entered (so as to define or redefine it).

• Select the tool which is desired to be made the master tool, say, tool number 1, by pressing the tool change button on the machine operator's panel required number of times. Alternatively, execute the appropriate tool change code (T0100 for tool number 1) in the MDI mode.

• Move this tool in the negative X direction to let it touch the workpiece at a *known* diameter position (say, X = 10) manually, using continuous feed (JOG mode), incremental feed (INC mode) or handle feed (HANDLE mode), as appropriate / convenient. For accuracy in positioning, it is recommended to use successively smaller increments (say, 0.1 mm, then 0.01 mm, followed by 0.001 mm), as the tool approaches the desired position. Continuous feed should be used only for quickly bringing the tool near this position, initially. To judge whether the tool has actually touched the surface, turn the spindle manually and look for a tool mark on the surface. Another method involves inserting and continuously shaking a thin piece of paper between the tool and the workpiece while the tool is made to approach the workpiece. The moment the paper gets stuck, the tool can be considered to have touched the surface. While inputting the values in the offset tables, allowance for the thickness of the paper can be taken, if felt necessary.

• Press OFS/SET key on the MDI panel. OFFSET, SETING, WORK and OPRT soft keys will appear. Pressing OFFSET soft key will display WEAR, GEOM and OPRT soft keys.

• For setting work offsets, press the WORK soft key. If it is already highlighted, i.e., selected by default, there is no need to press it again. The currently active workpiece coordinate system is displayed at the top left corner of the work offset screen. If the workpiece coordinate system to be edited is other than the active coordinate system, then the corresponding G code (say, G58) must be executed in the MDI mode to make it active. Now, bring the cursor to any coordinate display of the *active* WCS.

• Type X10 (for this example) and press MEASUR soft key. This will make the correct entry (overwrite the previous value) for the X component of the work offset.

• Now, retract the tool and then make it touch the face of the workpiece by moving it axially. Type Z0 and press the MEASUR soft key. This would enter the Z component of the work offset.

Note that the X0 position on a lathe is always along the axis of the chuck. It is, thus, independent of the length or the diameter of a workpiece. Therefore, once the X offset is correctly established, it need not be repeated in a subsequent offset setting process unless the position of the tool on the turret has been disturbed.

The method of defining WCSs on a milling machine is similar. A 3-axis milling machine has three components of work offset (X, Y and Z). For determining X and Y values, the spindle axis is required to be aligned with the desired XY datum. There is a difficulty in this because of the finite diameter of the tool. A simple and fairly accurate method is to use a sharp edged pencil type "tool" held in the spindle. Its pointed tip can be very accurately brought to the desired XY datum. For higher accuracy, the Z gap between the tip and the workpiece should be as small as practicable, and one should check the alignment by viewing it from two orthogonal directions so as to eliminate parallax. Thereafter, following the method adopted on a lathe, one needs to type, one by one, X0 and Y0 followed by the MEASUR soft key. For determining the Z offset, the master tool (which can be same as the pencil type tool) is made to touch Z0 (top surface of the workpiece) or a known Z level, for which a block of standard height kept over the workpiece can be used. Thereafter, typing the appropriate Z coordinate, followed by the MEASUR soft key, enters the Z component of work offset.

For determining the geometry offset of a desired tool on a lathe, bring it in the cutting position, and make it touch the surface and the face of the workpiece one by one. Press OFFSET soft key followed by GEOM soft key. On the geometry offset screen, bring the cursor to the appropriate row number (offset number), say, row number 2 for tool number 2, under any column, type appropriate coordinate (such as X10 or Z0) and press MEASUR. This should be done in the WCS appropriate for the workpiece held in the chuck. The nose radius and the tip number can be entered in the desired row by highlighting these one by one, typing the value and then pressing the INPUT soft key in any WCS.

The wear offsets for X, Z and R are initially kept zero, and later changed as per requirement. For convenience, INPUT+ soft key is also provided which enters the specified value incrementally, i.e., adds it to the existing value algebraically.

On a milling machine, the tool length offset can be entered in a similar way. We select the MDI mode, the tool for which length offset is to be determined is placed in the spindle, and the WCS, relevant for the workpiece on the table, is made active. Now the tool is made to touch the desired Z0 (or any other known Z position) surface. The geometry column of H in the desired row (offset number) is highlighted. Typing Z0 (or the appropriate Z coordinate), followed by the MEASUR soft key would enter the length offset which is to be used with G43 in the program. The radius of the tool can be entered in the geometry column of D, using the INPUT soft key. Wear values are initially kept zero, and edited suitably as and when required.

The effect of the entered offset values can be seen only after calling the tool with the appropriate offset number in the appropriate WCS. The coordinates would be displayed as

103

expected. Pressing the RESET key nullifies the tool offsets (which can be called again).

Once the tool offsets are determined for all the tools with respect to a master tool (on both lathe and milling machines), the process need not be repeated for a new WCS or when an old WCS needs to be redefined. We just need to use the *same* master tool for defining a new WCS or redefining an existing WCS. Minor inaccuracies, if any, because of any reason such wear of the master tool, can be taken care by suitably adjusting wear offsets of other tools. One should understand that tool offsets are the differences in the positions of different tools with respect to the master tool, which obviously is independent of the WCSs.

VERIFYING THE OFFSET SETTINGS

Let us assume that, on a lathe, the G58 work coordinate system has been used for a particular workpiece, and tool number 2 uses offset number 2. For verification, place the *same* workpiece in the chuck at its designated place, and execute

G58;
T0202;

in the MDI mode. T0202 must be executed even if tool number 2 is in the cutting position. Note that a control reset clears the offset adjustment for a tool (the display on the MDI screen would change to, say, T0200 from the previous T0202, and the coordinate display would accordingly change). Also, whenever the offset values are modified in some offset number (02, in the present example), its effect in the position display will not be reflected without executing T0202 *again*.

Now, manually bring the tool to a known position (in, say, JOG mode), and verify if the position display on the MDI screen is correct. If radius compensation is also being used, verify that the R and T columns of the geometry offset table show the correct radius and nose direction, respectively, in row number 2. The wear value of the radius (in the wear offset table) should be zero or close to zero (if modified, to take care of inaccuracies in radius compensation). Repeat the process for all the tools which are to be used.

On a milling machine, let us assume that tool number 2 is being used with offset number 2 in G58 for a particular workpiece. Place tool number 2 in the spindle, and bring the tip of the tool manually to a safe height (say, Z = 200 mm; do not worry if the screen does not display the correct position), and then execute the following in the MDI mode:

G21 G94 G58;
G01 G43 Z200 H02 F10;

A low feed has been deliberately used to get sufficient time to stop the machine in case the tool proceeds to a dangerous region due to offset errors. Since the tool is already at Z = 200 (approximately), not much movement is expected.

Now, bring the tool to a known position and verify if the position display on the screen is correct. If radius compensation is also being used, verify that row number 2 shows the correct radius in the D (geometry) column of the offset table. The wear value of the radius should be zero or close to zero in the D (wear) column.

METHODS OF OFFSET SETTING ON A LATHE

The method of offset setting on a lathe as described in Fig. 3 needs to be slightly modified to suit different tool geometries (It is assumed that all the tools are placed in the inverted position on a rear type lathe):

Right hand turning tool: As described in Fig 3.

Left hand turning tool: Z = 0 can only be approximately set, as the tool cannot touch the Z = 0 surface. If we have a partially machined workpiece (which will usually be the case for a left hand tool) on which a flat surface in the negative Z direction (i.e., a left face) is available, the tool may be made to touch this surface, and the corresponding Z value (negative, in this case), which can be measured by a vernier caliper, be specified.

Neutral turning tool: $Z = 0$ cannot be very accurately set. Bring the cutting edge of the tool approximately at the axial position of the right edge of the workpiece. The radial gap between the tool and the edge of the workpiece should be as small as possible, for a better accuracy. A magnifying glass may be used to verify the axial alignment more accurately.

Grooving/parting tool: These tools are not the single point cutting tools. The whole of the bottom edge cuts the material. However, the suggested method selects the lower left corner as the reference point, and the program is written for controlling the movement of this point. The programmer must know the width of the tool before writing a program, as he will have to take care of cutting by the whole of the bottom edge.

Threading tool: The method described for a neutral turning tool is used. However, accuracy in locating the $Z = 0$ plane is not very critical.

Drilling tool: The side of the drill bit is made to touch the outer surface of the workpiece, and at that moment, the diameter of the workpiece plus the diameter of the drill bit is specified as the X value. (We have assumed X axis drilling tool.)

Internal tools: The methods are the same as those for the external tools, except that, in this case, internal surfaces of known diameters are used for the X offset setting. A theoretical method can be to touch the outer surface from the opposite radial side of the workpiece, and use the corresponding diameter as the negative X value. This, however, is likely to give $-X$ overtravel alarm.

An alternate method: Note that in case of a neutral tool, an indirect method can correctly establish the Z reference. Let the left side of the tool shank touch the face of the workpiece. If the cross-section size of the shank is 25 mm × 25 mm, the current axial position of the tool tip is $Z = 12.5$ mm (because the insert is symmetrically placed on the shank). This method has to be used for a round insert with a large radius if the bottom of the insert is to be made the reference point, because the previous method may give unacceptable errors. This method can also be used for a left hand tool in a similar manner. If the tool tip is flush with the right side of the shank, it is at $Z = 25$ mm when the left side of the shank touches the face of the workpiece. In the case of neutral internal tools, the distance of the tool tip from the end face of the tool holder must be known (look into the tool handbook) for this method of Z reference setting.

VOLUME 5:

Understanding G32, G34, G76 and G92
on a
Fanuc Lathe

PREFACE

This is the fifth e-book in the series, "CNC Programming Skills." It explains the use of G32, G34, G92 and G96 for thread machining on an i-series Fanuc lathe or a similar control.

While information about these G codes are freely available in books and several online CNC forums, these are in scattered form, and many a time, not very clear. The present book is a comprehensive source of information about thread machining on a lathe. Apart from the usual descriptions, it also discusses some lesser known techniques such as thread machining using two tools at different RPMs.

This book is written keeping in view the requirements and expectations of beginners who wish to learn on their own with no helping hand available to them. No background is assumed. It even explains the thread nomenclature in the beginning. One only needs to have prior knowledge of basic codes such as G00 and G01 for following the book. Read on and find out yourself. Watch yourself growing into an expert programmer!

INTRODUCTION

Thread machining requires proper synchronization between the spindle rotation and the tool motion so as to ensure a uniform lead (which is same as the pitch for a single-start thread). Moreover, every subsequent pass of the tool should correspond to the same threading helix. Threading cycles are designed to automatically take care of these.

Two types of threading cycles are generally used (apart from G32 which is not a cycle; it is like a "modified" G01, fulfilling the threading requirements; and G34 is similar to G32 with a variable lead. These are described in the end):

• Single threading cycle, G92

• Multiple threading cycle, G76

The main difference between the two cycles is that whereas G92 needs to be called several times (once for each pass, with gradually increasing depth, till the final depth is obtained), G76 makes the complete thread in one call, automatically calculating the X levels of all the intermediate passes, apart from incorporating several other useful features. Although G92 does not provide so much flexibility as compared to G76, it is easier to use.

The value of F in the calling blocks of both the cycles is the lead of the thread. So, effectively, it is the *constant* feed per revolution, as if G99 is active, even though we may have actually used G98 (feed per minute) in the beginning of the program. G98/G99 have no effect on the behavior of threading cycles.

THREAD DIMENSIONS

A thread is specified by the outer diameter (OD) of a screw/bolt, which is the crest-to-crest diameter. All the dimensions of both internal and external threads, corresponding to an OD, are tabulated in handbooks, for standard threads. Dimensions of non-standard threads (i.e., those with non-standard OD and/or non-standard pitch) would need to be calculated. For example, the formulas for metric threads for a given OD and pitch (even for multi-start threads) are

Depth of thread = 0.61344 × Pitch

Core diameter of external thread = OD − 2 × Depth of thread

Bore diameter of internal thread = OD − Pitch

Root diameter of internal thread = Bore diameter + 2 × Depth of thread

For example, for a metric thread with OD 30 mm and pitch 1.5 mm (M30×1.5), the dimensions can be calculated as

Depth of thread = 0.61344 × 1.5 = 0.92016 mm

Core diameter = 30 − 2 × 0.92016 = 28.1597 mm

Bore diameter = 30 − 1.5 = 28.5 mm

Root diameter = 28.5 + 2 × 0.92016 = 30.3403 mm

The nomenclature of threads is shown in Fig. 1, with reference to the design profile of metric threads. It is the same for other thread forms also.

THREAD MACHINING

An external thread is machined from the OD to the core diameter (with the subsequent thread-cutting passes moving downwards, i.e., towards the axis of the workpiece), whereas an internal thread is machined from the bore diameter to the root diameter (with the subsequent thread-cutting passes moving upwards, i.e., away from the axis of the workpiece).

Sometimes, due to deformation caused by blunt threading-cutting edges or other reasons, the outer diameter of an external thread slightly increases after threading, resulting in mismatch with the corresponding nut. In such cases, the initial diameter of the stock may be reduced by 1-2 %. Similar adjustment might be needed in the bore diameter of an internal thread also (1-2 % increase in this case).

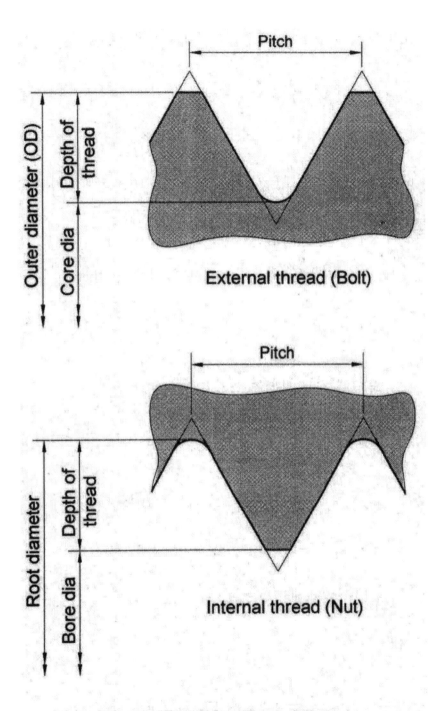

Fig. 1: Thread nomenclature

SINGLE THREADING CYCLE, G92

It belongs to G-code Group 1, along with G00, G01, G02, G03, G32, G34, G90 and G94. It is a modal code which remains effective till another code from the same group is commanded. The syntax is

G92 X(U)_ Z(W)_ R_ F_ Q_;

The meanings of the different words are explained below.

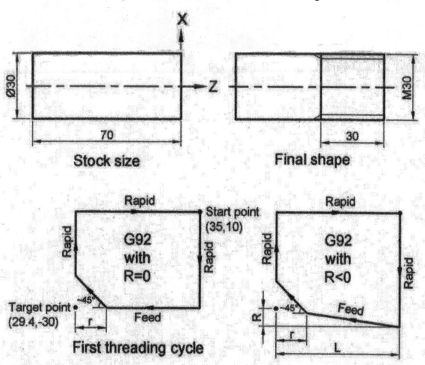

Note:
R (= L/32 for NPT or BSPT) is measured from the specified target point. Also, the value of R is kept same in all the subsequent G92 cycles; only the target point is gradually shifted to its final radial position.

Fig. 2: Single threading cycle, G92

116

Start and Target Points

$X(U)$ and $Z(W)$ are, respectively, the radial and the axial ends of the thread which can be viewed as the target point of the cycle (Refer to Fig. 2). U and W are the incremental coordinates of the target point with respect to the start point of the cycle. The position of the tool at the time of commanding G92 automatically becomes the start point of the cycle.

Taper Amount

R, which is a *radial* value, is used for making taper threads such as NPT (National Pipe Threads) and BSPT (British Standard Pipe Threads). If not specified, $R0$ is assumed which makes straight threads.

As shown in Fig. 2, the first rapid radial movement of the tool is up to the X coordinate defined by the X coordinate of the target point *plus* $2R$ (the factor of 2 comes because R is a radial value). This is a general statement which is always true irrespective of the position of the target point with respect to the start point. Therefore, R must be specified with proper sign so as to match the orientation of the taper.

Both NPT and BSPT pipe threads have the same taper, 1 in 32 on radius. Therefore, the value of R would be $L/32$ where L is the *axial* distance between the start point and the target point, as shown in Fig. 2. L is 40 mm in the example that follows, and R would be negative in this example, if tapered threads are to be produced.

Feedrate

F is the lead of the thread (Lead = Pitch × Number of thread-starts, e.g., the lead is double the pitch for a double-start thread). It is the axial displacement of the tool in one revolution of the spindle. It is, thus, equivalent to feed per revolution.

A pulse is generated in a particular angular position of the rotating spindle, once in every rotation. It is called 1-turn pulse or marker pulse. The control starts the feed motion in a threading cycle the moment it receives the 1-turn pulse. Therefore, in order to follow the same threading helix in all the subsequent thread-

117

cutting passes, it is necessary that the initial axial position in all the threading cycles be kept same (the tool as well as the RPM also should remain unchanged, as discussed later).

It is also possible to delay the start of the feed motion by a desired angle with respect to the orientation corresponding to the 1-turn pulse. This is done through the Q word. Q is the delay angle when the feed motion for thread cutting starts. Q can be specified as an *integer* value in the range 0 to 360000, corresponding to 0° and 360°, respectively. If not specified, 0° is assumed, as if $Q0$ is programmed, and the feed starts at the 1-turn pulse. Q is not a modal data. It must be specified in every block it is needed.

The valid range of F is

0.0001 to 500.0000 mm in metric mode, and

0.000001 to 9.999999 inch in imperial mode.

Decimal Digits for Lead

Most machines are set to use IS-B (Increment System B) which accepts three digits after decimal in mm mode and four digits after decimal in inch mode, for axial distances (e.g., X10.123 in mm mode and X10.1234 in inch mode). Extra digits, if any, are truncated, i.e., ignored by the machine. IS-C accepts one extra decimal digit, compared to IS-B, i.e., four in mm mode and five in inch mode. However, *irrespective of* the increment system being used on the machine, the lead of a thread can be specified with up to four decimal digits in mm mode and six decimal digits in inch mode, like what is allowed for feedrate in feed per revolution mode.

Allowable Feedrate

Depending on the combination of the control and the machine tool, accurate lead is obtained when the thread-cutting feedrate (Feed per minute = Lead × RPM) is less than a certain value, usually lying between 3000-4000 mm/min. If a higher feedrate is used, the tool will move at a slower speed, and the thread lead will be a bit finer. And, the machine will not generate any alarm! Therefore, if necessary, reduce the RPM to decrease the feedrate (which is 700 mm/min in this example).

Lead at Start/End of Thread

At any RPM, there will always be some inaccuracy in the thread lead at the start as well as at the end of a thread because of acceleration/deceleration in the feed motion which disturbs the synchronization between the spindle rotation and the tool motion. The length, over which this error (finer lead) occurs, increases with RPM. Therefore, use a lower RPM if sufficient run-in/run-out (also called, lead-in/lead-out) distances cannot be provided.

Multi-Start Threads

Q need not be specified for single-start threads, as it is not needed. However, it can be very advantageously used for machining a double-start thread which has two uniformly spaced independent helices. These can be made by choosing $Q0$ and $Q180000$, respectively, for the two helices. In this method, the start points for all the G92 cycles should be *same*. Similarly, a triple-start thread can be made by selecting $Q0$, $Q120000$ and $Q240000$ for the three helices, with the start points remaining at the same place every time.

Another method for machining multi-start threads involves shifting the start points axially by required amounts, without specifying Q (i.e., with $Q0$). This is explained later with reference to G76, the method being applicable to G92 also.

Chamfer Distance/Angle

The chamfer distance r depends on parameter 5130 which can be varied in the range 0 to 127 in the steps of 1. This, in turn, varies r in the range 0 to $12.7F$ in $0.1F$ increments.

The default setting of the chamfer angle is 45°, though it can be changed in the range 1° to 89° through parameter 5131 by specifying an *integer* value between 1 and 89. The actual chamfer angle would be slightly different than that selected, due to some lag in the servo system, though this is not important. Note that the tool never actually reaches the target point if a non-zero chamfer distance is specified.

Sample Program

In the example that follows, an M30 single-start external thread of 30 mm threaded length (without taper) from the right face is being made using G92, as shown in Fig. 2. The pitch and the core diameter are 3.5 mm and 25.706 mm, respectively, which can be read from a metric threading chart.

O0001; (Program number 1)

G21 G97 G98; (mm mode, constant RPM, and feed in mm/min selected. Note that the constant surface speed, G96, cannot be used in threading, as a change in the RPM causes a shift in the threading helix which would spoil the thread)

G54; (Workpiece coordinate system)

G28 U0 W0; (Slides go to the reference position)

T0101; (The turret rotates to bring the tool number 1 with offset number 1 in the cutting position. It is assumed that tool number 1 is the threading tool)

G00 X35 Z10; (This position becomes the start point of G92. About 2-3 thread-lead run-in, i.e., the axial clearance between the start point and the workpiece, 10 mm taken here, must be provided to accommodate the initial error in the lead caused by the finite acceleration when the feed motion starts; otherwise, the first few threads would have incorrect lead. The inaccuracy occurs over a larger length at higher RPMs, requiring more clearance. Radial clearance, 2.5 mm taken here, is not critical with G92, but it should be more than the height of the threads if being machined by G76; otherwise, there would be interference when the tool retracts in feed-hold condition)

M03 S200; (CW RPM = 200 starts)

M08; (Coolant starts. M08/M09 might be machine-specific, as designed by the machine-tool builder)

G92 X29.4 Z–30 F3.5; (Or, G92 U–5.6 W–40 F3.5; The target point is X29.4 Z–30, and the lead is 3.5 mm)

X28.8; (Or, U–6.2; The second threading cycle. Modal G codes, modal data as well as unchanged axis addresses need not be repeated in subsequent blocks, if their use is again intended. Therefore, this block actually means G92 X28.8 Z–30 F3.5)

X28.3; (The third threading cycle)

X27.8; (The fourth threading cycle)

X27.4; (The fifth threading cycle)

X27.0; (The sixth threading cycle)

X26.7; (The seventh threading cycle)

X26.4; (The eighth threading cycle)

X26.2; (The ninth threading cycle)

X26.0; (The tenth threading cycle)

X25.9; (The eleventh threading cycle)

X25.8; (The twelfth threading cycle)

X25.75; (The thirteenth threading cycle)

X25.706; (The last threading cycle. 25.706 mm is the core diameter)

X25.706; (A spring pass, to take care of any spring-back in the workpiece)

M09; (Coolant stops)

M05; (The spindle stops)

G28 U0 W0; (The slides go back to the reference position)

M30; (Causes control reset and program rewind)

Depth of Cut

Because of the V shape of the threading tool (for a metric or a similar thread), the width of the chip goes on increasing in subsequent threading passes. This means that if the depth of cut is kept same in all passes, the amount of material removed in a pass would go on increasing. This would result in a progressive increase in the cutting forces which is not desirable. Therefore, it is better to start with a large depth of cut and *gradually* decrease it to a small value. In the given program, it has been varied from 0.3 mm down to 0.022 mm. It involves manual calculations which is tedious and may not be perfect. This is a major shortcoming in G92. G76, discussed next, does this appropriately and automatically.

Feed/RPM Overrides

The required feedrate during threading (G32, G34, G76 or G92) is calculated and used by the machine control, based on the specified values of the lead and the RPM. The feed-override switch and the RPM-override switch both automatically becomes ineffective and remain fixed at 100%. The reason for disabling the override switches is that even though the control would adjust the RPM/feedrate appropriately when the other is changed, so as to synchronize their motions, it cannot maintain the *same* threading helix which would spoil the threads. In fact, because of this reason, G96 should not be used during threading, as stated earlier.

Using Two Tools

If the threading job is extensive, the threading tool would wear out very soon. In such a case, it might be more economical to use two different threading tools—one for roughing and the other for finishing. Even a slightly worn-out tool can be used as a roughing tool which should be used for all the threading passes except the last 2-3 passes, for which a finishing tool may be employed. Even a different RPM for the finishing tool might be used, if desirable. This would reduce the cost of tooling to a great extent, and at the same time, the quality of machining would also improve.

Shift in Threading Helix

A serious problem with using two tools is that, even if the start points are kept same, once the RPM and/or the tool is changed, the machine starts following a different threading helix which will spoil the thread. To overcome this problem, the operator will have to experimentally determine the axial shift of the helix when the second tool is used, and add/subtract (as appropriate) this amount to/from the initial axial position of the first tool, to determine the correct axial start position of the second tool. For example, if the first tool starts machining from $Z = 10$, and the helix of the second tool lies 1 mm to the left of the helix of the first tool, the start position of the second tool should be changed to $Z = 11$, to make the two helices coincide. The amount and the direction of the shift (left or right with respect to the helix formed by the first tool) can be found out by conducting a simple experiment, described next.

Measuring Shift in Helix

Consider the case of an M30 thread, for example. Take a workpiece of 30 mm diameter and run the following program with the first tool:

G00 X35 Z10;

M03 S200;

G92 X29.9 Z–10 F3.5;

This will just scratch a thin helix on the workpiece. Apply some color to it to differentiate it from the second helix to be made next. Now, run the same program with the second tool, without unclamping/re-clamping the workpiece. Another helix will be made. Using some accurate measuring device, such as a tool-makers' microscope, measure the shift between the two helices, and also observe the direction of the shift.

Note that the experiment described above will need to be repeated for every outer diameter and every RPM. The RPM during the experiment must be the same as that in the original program. However, the RPMs for the two tools can be chosen to be different, if desired. Also, the entire threading operation, i.e., roughing and finishing, both should be done in the same setup. Once a partially threaded workpiece is unclamped and re-clamped, the same helix is not guaranteed in further machining. This also means that rework on a thread is not possible using conventional methods, once it is unclamped.

Taper Caused by Flexing

Sometimes, when the length-to-diameter ratio is large, the workpiece starts flexing during threading. In fact, if the OD is to be made by turning a larger-diameter stock with a single-point turning tool, this problem would again be there. This results in taper threads, with the diameter at the right end being larger. If special tools/attachments are not available, and one must use single-point tools, then a workable solution to this problem can be to use the method of taper turning/threading, in the opposite direction, so as to compensate for the erroneous taper. Essentially, one would need to estimate the taper obtained (on radius), and use it with R with a negative sign.

Another way to take care of flexing is to turn, say, half the length to the outer diameter and make the threads over it. Thereafter, turn the remaining part, and thread it. If required, the length can be divided into even three or more parts. To make sure that the threading helix is the same every time, it is necessary that the same start point and the same RPM be used for *all* threading cycles. There would, of course, be threading cutting in air for the already threaded length, but there is nothing one can do about it. This is, however, not a very serious problem because threading feedrate is rather high, because of which there would not be much effect on the overall cycle time. The author thankfully acknowledges that this idea was taken from Post#2 of a thread at practicalmachinist.com forum,

http://www.practicalmachinist.com/vb/cnc-machining/cnc-lathe-small-od-threads-315775/.

A serious problem in the method of making a thread in multiple segments, as described above, is the error in the lead at the end of each thread segment, because of the inherent deceleration at the end of the feed motion. A simple solution to this problem is to provide enough chamfer distance so that the part of the thread with the incorrect lead is cut in the air (Post#12 of the same thread).

Machining Internal Threads

For making, say, an M30 internal thread, first a hole of bore diameter 26.5 mm would need to be drilled. The start point for G92 can be taken at X21.5 Z10. Thereafter, machining would be done in several passes to finally reach the root diameter 30.794 mm. These values can either be read from the metric threading chart, or calculated using the given formulas.

Limitations of G92

Though quite simple to use, the single threading cycle, G92, suffers from the following limitations:

• It needs to be called a number of times corresponding to each cutting depth.

• The programmer has to use his own judgment regarding the depth of cut for each pass, which should be gradually decreased so as to maintain nearly same cutting force in all passes.

• Both the edges of the cutting tool cut the material, which increases the load on the tool tip, especially during the final passes.

• The chamfer distance cannot be programmed. Parameter 5130 needs to be manipulated for the desired chamfer distance.

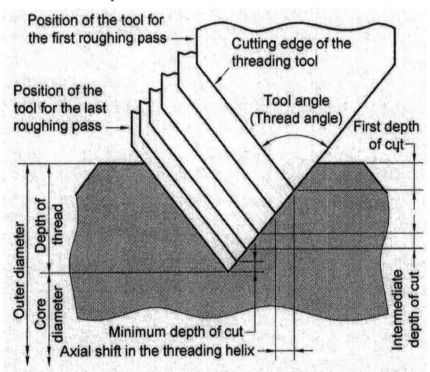

Note: Finishing passes not shown (Zero finish allowance assumed)

Fig. 3: Axial shifts in the subsequent roughing helices of G76

125

MULTIPLE THREADING CYCLE, G76

G76, the multiple threading cycle, has none of the limitations of G92. In addition, it has several special features:

• Just one call makes the entire thread.

• The first depth of cut, specified in the second block of G76, locates the first roughing pass. Thereafter, the depth of cuts for the subsequent passes are *calculated* by the control. These are gradually decreased so as to ensure equal material removal in every pass (Fig. 3). This keeps the cutting forces nearly constant in all the passes. If the calculated depth of cut comes out to be smaller than the minimum depth of cut, specified in the first block of G76, the subsequent depth of cuts remain fixed at the specified minimum value.

• There is also a provision to include some (01 to 99) finishing passes which machine the material left as finishing allowance, specified in the first block of the cycle. Unlike the roughing passes, there is no shift in the threading helix during the finishing passes, and hence, both the edges of the tool cut the material, which finishes both sides of the thread groove. All the finishing passes are executed at the same radial level (at the final X level). Therefore, only the first finishing pass is the cutting pass; the remaining finishing passes are spring passes, provided to remove any un-cut material, left because of a possible flexing in the workpiece which is sometimes observed while threading a slender workpiece without using the tailstock. The cycle is shown in Fig. 4.

• Major machining (roughing) is done by only one edge of the threading tool. To ensure this, the control automatically shifts every subsequent threading helix axially in the direction of feed by $d \tan \theta$ where d is the current depth of cut and θ is half of the tool angle (this is the reason why the tool angle needs to be specified in the G76 syntax). However, in finishing passes, the helix is not shifted, as mentioned earlier. The feature of shifting the helix, however, results in one edge of the threading tool wearing out more than the other edge. To avoid this, G76 should be used for making threads from left to right also, wherever possible.

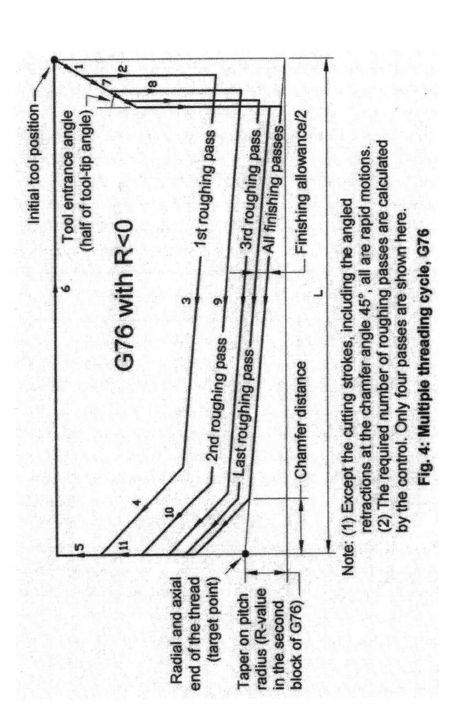

Fig. 4: Multiple threading cycle, G76

Note: (1) Except the cutting strokes, including the angled retractions at the chamfer angle 45°, all are rapid motions.
(2) The required number of roughing passes are calculated by the control. Only four passes are shown here.

127

Syntax

It is a two-block canned cycle belonging to G-code Group 0, along with other non-modal (one-shot) codes such as G70–G75 (The one-block G76 format is not very popular because of certain limitations. The selection between the two is through parameter 0001#1(FCV); 0 for two-block format, and 1 for one-block format):

G76 P_ _ _ _ _ _ Q_ R_;

G76 X(U)_ Z(W)_ P_ Q_ R_ F_;

The first two digits of P (in the first block) are the number of finishing passes (01 to 99 allowed).

The middle two digits of P, divided by 10 and multiplied by the lead, is the chamfer distance (00 to 99 allowed). Parameter 5130, described earlier, automatically gets overwritten by the specified value. Therefore, if a G92 cycle is executed subsequently, it would use the modified chamfer distance.

The last two digits of P is the angle of the thread (80°, 60°, 55°, 30°, 29° or 0° can be selected).

Q in the first block is the minimum depth of cut in microns (in mm mode), beyond which the depth of cut is not decreased even if the value calculated by the control comes out to be smaller than this.

R in the first block is the finishing allowance on *diameter*.

$X(U)$ $Z(W)$ is the target point, i.e., the radial and the axial end of the thread, as with G92. Thus, X is the core diameter of an external thread, and the root diameter of an internal thread.

P is the height of the thread in microns (in mm mode) which can be read from threading charts, corresponding to the given outer diameter of the thread.

Q in the second block is the first depth of cut in microns (in mm mode) which applies to the first roughing pass only.

R in the second block is the taper amount on radius, as with G92.

F is the lead of the thread (= Pitch × Number of starts)

Least-Input-Increment Data

Q in the first block, and both P and Q in the second block do not accept decimal digits. These are specified in multiples of the least input increment being used on the machine, without a decimal point. Thus, in millimeter mode, micron (0.001 mm) is used, and in inch mode, ten-thousandth of an inch (i.e., 0.0001 inch) is used, for the commonly used increment system, IS-B. For example, for 0.5 mm, Q500 is programmed in millimeter mode. In inch mode, for 0.05 inch, Q0500 (or simply Q500) is programmed. Leading zero(es) can be omitted.

Roughing/Finishing Passes

The first depth of cut, which locates the first roughing pass, is measured from the outer diameter (bore diameter, in the case of an internal thread). The control calculates the outer diameter by adding twice the height of the threads to the core diameter. Bore diameter is calculated by subtracting twice the height of the threads from the root diameter. This is the reason why the height of the thread is required to be specified in this cycle. The last roughing pass is at the level of the core diameter plus the finishing allowance (root diameter minus the finishing allowance, in the case of an internal thread). The total number of roughing passes is decided by the control, based on the height of the thread, and the first and the minimum depth of cuts. The programmer does not have to worry about it, and in fact, he has no control over it. He can only specify the number of finishing passes.

Roughing-Passes Formula

In order to ensure equal volume removal in each pass the control calculates the depth of cut for n^{th} pass using the formula

$$d_n = d_1\sqrt{n}$$

where d_1 is the first depth of cut (Q in the second block of G76).

This formula assumes that the tool angle is 60° and the tip radius is zero. Therefore, it is only approximate, but works well. In any case, we cannot change the formula.

Taper Threads

It is also possible to make a taper thread by specifying an R value, with proper sign, in the second block of G76. The meaning and the sign convention for R is the same as that in the G92 cycle, described earlier. The reference point for measuring R is the radial and axial end of the thread, specified as the target point in the second block of G76. R is always calculated on the basis of the start point (the position of the tool just before calling G92 or G76) and the target point. Features like chamfer distance and shift in threading helix do not affect the calculations for R. Calculate R as if you are using G92, considering the *total* axial distance between the start point and the target point (marked L in Fig. 4). Thus, for NPT or BSPT pipe threads, R would be $L/32$, with proper sign, as in the case of G92.

Square Threads

For square threads, the tool angle is zero. In such a case, the helix is not shifted for subsequent passes, and the cycle runs like the single threading cycle, except that the depth of cut gradually decreases, as explained earlier. The width of the tool, which is similar to a grooving tool, should be equal to the width of a single thread (i.e., half of the thread pitch).

Buttress Threads

A buttress thread has unequal flank angles. Both the American-standard and the British-standard buttress threads have 45° leading-edge angle and 7° trailing-edge angle. The German standard has 33° and 3° angles, respectively. Some other types such as 45°/5° and 45°/0° are also in use. The terms "leading-edge" and "trailing-edge" refer to the orientation of the threading insert with respect to the feed direction.

For a thread, where the leading and the trailing angles are different, twice the *trailing angle* is specified as the angle of the thread. This causes the insert to slide down the trailing edge, while the major cutting is done by the leading edge. Accordingly, if the trailing angle is 7°, then the angle specified in the threading cycle would be 14°. This angle, however, is not permitted in the two-block G76 format. In such cases, the nearest smaller angle should be specified, which is 0° in this case. If the leading/trailing edges are reversed, 80° would be the most appropriate value. When the specified angle in the G76 cycle is smaller than twice the trailing angle, then some cutting would take place also by the trailing edge of the insert.

Constant-RPM Requirement

As mentioned previously, the spindle RPM must not be changed in the middle of a threading cycle. So, G96 cannot be used in the program. The feed override and the RPM override switch automatically become ineffective and remain fixed at 100% during threading.

Servo-Lag Effect

As in the case of G92, a run-in of about 2-3 pitch widths is required at the start of the thread, otherwise the start of the thread may have an incorrect pitch. Similarly, the last few threads may have incorrect pitch. This happens due to the inherent lag in the servo system during acceleration and deceleration. To account for this, sufficient run-in in the beginning as well as run-out in the end (say, inside a groove) should be provided. The problem is less severe at lower RPMs.

Sample Program

The following program makes the same thread which was made by G92 previously:

O0002;

G21 G97 G98;

G54;

G28 U0 W0;

T0707;

G00 X35 Z10;

M03 S200;

M08;

G76 P031560 Q150 R0.15;

G76 X25.706 Z–30 P2147 Q250 F3.5;

M09;

M05;

G28 U0 W0;

M30;

The threading cycle in this program is based on the following data:

Number of finishing passes = 03

Chamfer distance = $(15 \div 10) \times 3.5 = 5.25$ mm

Thread angle (tool-tip angle) = 60°

Minimum depth of cut = 150 micron = 0.15 mm

Finishing allowance (on diameter) = 0.15 mm

Core diameter = 25.706 mm

Axial end of thread = 30 mm in the negative Z direction

Depth of thread = 2147 micron = 2.147 mm

First depth of cut = 250 micron = 0.25 mm

Lead (= pitch, for single-start) = 3.5 mm

Roughing and Finishing Passes

The control gradually decreases the depth of cut, starting from 0.25 mm, so as to ensure equal volume removal in every pass. If the calculated depth of cut turns out to be smaller than 0.15 mm, the depth of cut for the remaining roughing passes is taken as 0.15 mm. The roughing passes end at X25.856, leaving the specified finishing allowance (R0.15). Thereafter, the finishing passes start, the first with a depth of cut of 0.075 mm, and the remaining two exactly re-tracing the first finishing pass, as spring passes.

Multi-Start Threads

If a multi-start thread (say, a double-start thread) is to be machined, two evenly-spaced threads will have to be made separately. For both the threads, the *F* value would be twice the required pitch (i.e., equal to lead), and the starting *Z* location of the second thread would need to be shifted by the desired pitch distance (half of the *F* value). This technique works for both G92 and G76. For example, the following changes are required in the previous program, for a double-start M30 thread:

G00 X35 Z10;

G76 P031560 Q150 R0.15;

G76 X25.706 Z–30 P2147 Q250 F7.0; *(F* = Lead = 2 × Pitch)

G00 X35 Z13.5; *(Z* = Previous *Z* + Pitch)

G76 P031560 Q150 R0.15;

G76 X25.706 Z–30 P2147 Q250 F7.0;

For a triple-start thread, three separate threads are required to be made:

G00 X35 Z10;

G76 P031560 Q150 R0.15;

G76 X25.706 Z–30 P2147 Q250 F10.5; *(F* = Lead = 3 × Pitch)

G00 X35 Z13.5; *(Z* = Previous *Z* + Pitch)

G76 P031560 Q150 R0.15;

G76 X25.706 Z–30 P2147 Q250 F10.5;

G00 X35 Z17; *(Z* = Previous *Z* + Pitch)

G76 P031560 Q150 R0.15;

G76 X25.706 Z–30 P2147 Q250 F10.5;

If G92 is desired to be used for multi-start threads, similar changes in *F* and the starting *Z* position would be required, without specifying *Q*.

Using Two Tools

The technique described previously in the context of G92, for using different tools/RPM for roughing and finishing operations, cannot be used in a similar manner in the case of G76. It is because the control shifts the threading helix for every subsequent roughing pass in this cycle. Even if we decide to use G92 for finishing, with a view to reduce the complication, we need to know the axial position of the last helix of G76, not of the helix on the surface of the job. However, one can make use of the fact that the last helix lies at a distance of $a \tan \theta$ to the left (or right, depending on the direction of the feed) of the helix on the *surface* of the job, where

θ is the half of the angle of the thread, and

a = (Height of thread – Finishing allowance/2)

Therefore, the total leftward shift in the helix of G76, with respect to the helix of G92, will be equal to the measured shift of the helix of G76 towards left (use a negative sign if the shift is towards right) on the surface of the workpiece plus (minus, if the feed is towards right) $a \tan \theta$. The initial tool position for G76 should be to the right of the initial position for G92 by this amount (if it is negative, the initial position for G76 should be towards left by this amount).

Finishing Allowance in G76

In the method described above, the finishing allowance in G76 may be specified as zero because the finishing is actually being done by G92, subsequently. With zero finishing allowance, if the number of finishing passes is specified as 1 in G76, which is the minimum permissible value, then the tool would machine at the level of the core diameter / root diameter twice at the end of the cycle, once as the last roughing pass and the second time as the sole finishing pass with zero finishing allowance.

Machining Allowance for G92

Because the finishing is to be done by G92, G76 should leave some machining allowance on core diameter (root diameter in the case of an internal thread) in the end. In other words, it should not machine entirely up to the core/root diameter. For this, the core diameter in the second block of G76 would need to be increased by the desired finishing amount (on diameter). For internal threads, the root diameter would be decreased by this amount. Since the outer diameter (bore diameter, for an internal thread) should remain unchanged (which the control calculates on the basis of the core/root diameter and the height of the thread), the height of the thread, specified in the second block of G76, should be decreased by the same finishing amount (on radius).

Helix on OD for G76

To obtain a helix on the surface of the workpiece, the first depth of cut of G76 should be very small, say 0.05 mm, and the execution of the cycle should be aborted after the first roughing pass is complete. To avoid any logical error in the execution of the cycle, the minimum depth of cut should be kept smaller than the first depth of cut.

Program for Combined G76/G92

Consider, for example, the same M30 thread which would now be machined first by G76, followed by G92. Following the experimental procedure described earlier, let us assume that the start point for G92 would need to be shifted by 2 mm to the right of the start point for G76. Also assume that G92 would be commanded twice, to remove 0.05 mm (radial value) in each pass, totaling to 0.1 mm on radius and 0.2 mm on diameter. Therefore, the core diameter for G76 would be 25.906 mm (= 25.706 + 0.2), and the height of the thread would be 2.047 mm (2.147 − 0.1). The program can be written as

O0003;

G21 G97 G98;

G54;

G28 U0 W0;

T0707; (Roughing tool)

G00 X35 Z10;

M03 S200; (Roughing RPM)

M08;

G76 P011560 Q150 R0; (One finishing pass and zero finishing allowance specified)

G76 X25.906 Z–30 P2047 Q250 F3.5; (Core diameter increased and height of thread decreased so as to maintain the same outer diameter. This leaves some machining allowance for G92; 0.2 mm on diameter, i.e., 0.1 mm on radius, in this case)

M09;

M05;

G28 U0 W0;

T0909; (Finishing tool)

G00 X35 Z12; (2 mm axial shift, as determined experimentally)

M03 S300; (Finishing RPM can be chosen to be different)

M08;

G92 X25.806 Z−30 F3.5; (First pass of G92 with depth of cut 0.05 mm)

X25.706; (Second pass of G92 with depth of cut 0.05 mm)

X25.706; (Spring pass of G92, if desired)

M09;

M05;

G28 U0 W0;

M30;

THREAD CUTTING WITH G32

Threading can also be done using G32. It belongs to G-code Group 1, along with G00, G01, G02, G03, G34, G90, G92 and G94. It is a modal code which remains effective till another code from the same group is commanded. Its syntax and the toolpath are similar to those of G01, except that the F word in G32 is the lead of the thread (specifying the feed per revolution), and it has an extra (optional) Q word, as with G92:

G32 X_ Z_ F_ Q_;

The lead is measured along the Z axis if the path of G32 makes an angle smaller than 45° with the Z axis. Therefore, in all the usual threading applications including taper threading, G32 calculates and maintains the required feed per minute *along the Z axis*, for the chosen RPM and the specified lead, using the relation

Feed per minute along the Z axis = Lead × RPM

Like in other threading G codes, the feed and the RPM overrides are automatically disabled to fulfill the threading requirements. And, G97 has to be necessarily used.

Unlike G76 and G92, G32 is not a cycle. Therefore, for machining a thread, a cycle will have to be simulated by inserting three G00 statements for retraction and repositioning the tool for the next pass. For example, an M30 external thread would require the following lines:

G00 X35 Z10;

X29.5;

M03 S200;

M08;

G32 Z–30 F3.5; (First pass)

G00 X35;

Z10;

X29;

G32 Z–30 F3.5; (Second pass) etc.

Like other threading cycles, G32 starts the feed after receiving the 1-turn pulse from the spindle encoder which generates one such pulse per revolution at a particular spindle orientation. This ensures the same threading helix in every pass.

Taper Thread Cutting

There is nothing special about it. One has to only provide both X and Z words in the G32 block to match the taper. For example, for taper pipe threads, the difference in the X coordinates of the start point and the end point (target point) should be equal to $L/16$ where L is the axial distance between the start point and the end point.

Tapping

Though it is not very convenient to use G32 for making an external thread, it is well-suited for tapping, for which G76 and G92 are not suitable. For example, a right-hand M16 tapping in a Φ14 mm hole (bore diameter) can be done in the following manner:

G00 X0 Z10;

M03 S200;

M08;

G32 Z–30 F2;

M05;

G04 P2000;

M04;

G32 Z10 F2;

Spindle stop (M05) and dwell (G04) before reversing the RPM, reduces the load on the spindle motor. The only problem with this program is that there would be a mismatch in the feed and the RPM at the axial end of the threads (i.e., at $Z = -30$), which will break the tap. This necessitates the use of a floating (spring-loaded) tap holder which permits some axial play. Some people use about 95% feed (F1.9, in the present example) in the forward pass and 105% (F2.1, in this example) in the reverse pass, which pulls out the floating tap a little because the threads of the tap force it to follow the correct feed. This reduces vibration, resulting in better surface finish. Floating tap holders are actually designed to operate in slightly stretched condition.

Continuous Threading

If there is a requirement to abruptly change the lead and/or the taper angle of a thread somewhere along the length, at one or more places, such that the same threading helix is maintained, this is not possible through G76 or G92. This can only be done through G32 which *does not wait* for the 1-turn signal if the *preceding* block is a G32 block. This maintains the same continuous helix even though the different thread segments, with different leads and/or taper angles, are machined by different (successive) G32 blocks.

After a pass is complete, and the tool is retracted for the next pass at a lower depth, the *first* (only) G32, in the next series of the G32 blocks, again looks for the 1-turn signal because its preceding block was not a G32 block. This repeats machining along the same previous helix. Thus, continuous threading in a single helix can be done.

In the example given below, which has two thread segments, the *X* words in the G32 blocks would not be there for straight threads:

G00 X_ Z_; (Initial position for the first pass)

G32 X_ Z_ F_; (The first thread segment. 1-turn signal is looked for because the previous block was not G32)

X_ Z_ F_; (The second thread segment. 1-turn signal is not looked for because the previous block was G32)

G00 X_; (Radial retraction)

Z_; (Axial retraction to the same initial Z)

X_; (Initial position for the second pass)

G32 X_ Z_ F_; (1-turn signal is looked for because the previous block was not G32)

X_ Z_ F_; (1-turn signal is not looked for because the previous block was G32) etc.

Multi-Start Threading

G32 allows the use of a *Q* word in its calling block. This can be used for delaying the threading-start angle appropriately, for machining along different helices. The principle is exactly the same as that discussed in the context of G92. The format of a program for a double-start external thread without taper is given below. The Z coordinate of the start points for all the G32 blocks is required to be kept *same*; only the X coordinate is gradually shifted downward for subsequent passes, till the core diameter is reached. This sequence needs to be repeated twice for a double-start thread: once with Q0 (which need not be explicitly mentioned, as its absence is interpreted as Q0) and the other with Q180000. For a triple-start thread, the sequence would need to be repeated thrice, with Q0, Q120000 and Q240000, respectively.

G00 X_ Z_; (Start point for G32)

G32 Z_ F_ Q0; (The first pass of the first helix)

<Rapid retraction to the start point for the second pass>

G32 Z_ F_ Q0; (The second pass of the first helix)

<Rapid retraction to the start point for the third pass>

.

.

<Rapid retraction to the start point for the last pass>

G32 Z_ F_ Q0; (The last pass of the first helix. Machining over for the first thread)

<Rapid retraction to the start point for the first pass of the second helix>

G32 Z_ F_ Q180000; (The first pass of the second helix)

<Rapid retraction to the start point for the second pass>

G32 Z_ F_ Q180000; (The second pass of the second helix)

<Rapid retraction to the start point for the third pass>

.

.

<Rapid retraction to the start point for the last pass>

G32 Z_ F_ Q180000; (The last pass of the second helix. Machining over for the second thread)

VARIABLE-LEAD THREADING WITH G34

G32 can be used for producing threads with a constant lead, or when the lead changes *in steps* along the length of the screw. It cannot produce a thread with a *uniformly-varying* lead. G34 has been designed for this requirement. It is similar to G32 in all respects except that it has an extra *K* word for specifying the change in the lead per revolution of the screw. The lead increases in the direction of the tool movement if the specified value is positive, and decreases if the value is negative. The *F* word is the lead at the *start* of the thread feed. The syntax is

G34 X_ Z_ F_ K_ Q_;

The valid range of *K* is

±0.0001 to ±500.0000 mm/rev in metric mode, and

±0.000001 to ±9.999999 inch/rev in imperial mode.

An unreasonable value of *K* may result in a negative lead, or the lead may exceed its maximum permissible value. This makes the machine produce P/S alarm number 14.

Example: If the lead at the start of the feed motion is 2 mm, and it increments at the rate of 0.1 mm/rev, the code would be

G34 Z_ F2 K0.1;

VOLUME 6:

Understanding G71 and G72
on a
Fanuc Lathe

PREFACE

It is becoming a trend to use CAM softwares for all programming applications. While it is true that one cannot do without these on multi-axis machines, these are not really needed on a two-axis lathe. If one knows how to use roughing cycles, grooving cycles and threading cycles, writing a program for a conventional part is a matter of minutes, for which the elaborate procedure of a CAM software would generally take a lot more time. Therefore, even if one has easy access to a CAM software, its use should be avoided for simple jobs. After all, who uses a calculator for adding a single-digit number to another single-digit number!

The built-in canned cycles on Fanuc lathes are so powerful and offer so much flexibility that once you start using these, you would never think of using any CAM software for conventional jobs. The present text, which is the sixth in the series "CNC Programming Skill," explains in detail the multiple rough-turning cycle (G71) and the multiple rough-facing cycle (G72) on an i-series Fanuc control using G-code System A. It is expected that the reader is aware of basic G codes such as G00 and G01, and basic M codes such as M03 and M30. One week-end should be sufficient to master these cycles, without referring to any other book/manual or seeking guidance from an expert.

MULTIPLE ROUGH-TURNING CYCLE, G71

Introduction

The multiple rough-turning cycle, G71, which accomplishes stock removal in turning, has two versions: Type-1, the earlier version, and Type-2, the later version. With Type-2 available on a machine, there is no good reason to use Type-1 which imposes a limitation to use only monotonically varying surface profiles (explained later). Moreover, even for the same profile (which is acceptable to Type-1), Type-2 gives a better finish. However, all the machines do not have Type-2, on which one has to use Type-1 only. The syntax of G71 for the two types is exactly same; only the method of defining the first point on the surface profile is different.

Syntax of G71

G71 has two formats: one-block format and two-block format. In the one-block format, the depth of cut and the radial retraction need to be specified through parameters 5132 and 5133, respectively, whereas these can be directly programmed in the two-block format. Thus, the two-block format offers more flexibility to the programmer for defining the required cutting conditions.

While it is possible to make a selection between the two formats of G71 through parameter 0001#1 (FCV), which can be set equal to 1 for the one-block format and equal to 0 for the two-block format, most machines are set for the two-block format only. It is also possible to manipulate this parameter by changing the PROGRAM FORMAT on the SETTING screen (MDI mode > OFS/SET key > SETTING soft key). In fact, all canned cycles, G71 through G76, have both one-block and two-block formats, which are selected by this parameter in the same manner. The syntax of the two-block format of G71 is

G71 U_ R_;
G71 P_ Q_ U_ W_ F_ S_ T_;

where

U in the first block is the depth of cut in roughing;

R is the radial retraction after each roughing pass;

P is the block number defining the first point on the profile;

Q is the block number defining the last point on the profile;

U in the second block is the X finishing allowance (on diameter), with sign;

W is the Z finishing allowance, with sign;

F is the feedrate in per minute or per revolution corresponding to G98/G99;

S is the RPM or CSS, corresponding to G97/G96;

T is the tool number and the offset number.

Cutting Parameters for G71

The F, S and T values, if specified while defining the profile, i.e., the values specified between the P block and the Q block, are ignored by G71. These are later used by G70 in the finishing operation where the specified finishing allowances are removed. G71 uses the F, S and T values specified in its own block. If a value(s) is not specified in its own block, the previously (last) specified value is used. Note that, the only purpose of T is to change the tool-offset number, if desired. The tool must not be changed, as changing the tool at an unsafe position may cause a crash.

Finishing Allowances

A roughing cycle such as G71, which is used with a large depth of cut and high feedrate (within the capacity of the machine) so as to maximize the productivity, cannot be expected to give very good accuracy and surface finish. Therefore, in G71 as well as in G72 and G73, there is a provision to specify finishing allowances along both radial and axial directions. A positive/negative finishing allowance shifts the defined profile along the corresponding axis in the positive/negative direction by the specified amount. The two finishing allowances can be positive, negative, or zero, in any combination, to suit the given profile. A roughing cycle is designed to produce the shifted profile, rather than the original profile. Thus, it leaves some extra material in the end, which is accurately machined by a subsequent finishing cycle G70, with maybe a different tool/feedrate/RPM/CSS, for better accuracy as well as surface finish.

In the subsequent discussion, we would refer to the original profile as the defined profile (because we define the original profile only), and the shifted profile as the desired profile (because this is what we want the roughing cycles to make). In Figs. 1 and 2, *PQ* is the defined profile, and *CD* is the desired profile. The shape of the profiles shown in Figs. 1 and 2 is arbitrary for illustration. A profile can only have linear and circular segments.

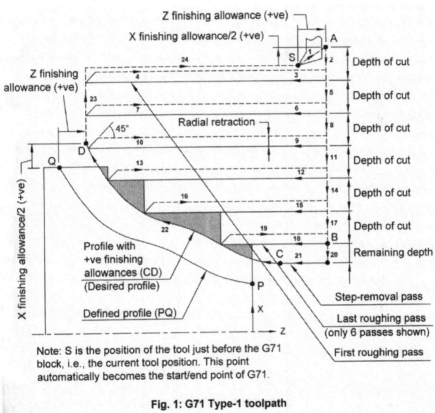

Fig. 1: G71 Type-1 toolpath

Selection between G71 Type-1 and Type-2

In G71 Type-1 and Type-2, the methods of defining the first point on the profile of the job are different. In fact, if both the types are available on a machine, the control selects between Type-1 and

149

Type-2, based on how the first point on the profile has been defined.

If the first point (point P in Figs. 1 and 2) is defined using *only* X or U, then Type-1 is selected by the control. In such a case, the axial position of the first point is *assumed* to be the same as that of the tool just before commanding G71, i.e., at the current axial position of the tool (Z coordinate of point S in Fig. 1). Since some axial clearance is generally provided for machining by G71 (typically, 2 mm), the first point on the profile and the first point assumed by the control would not be at the same axial position, though these are at the same radial position (because the X coordinate of the first point on the profile is used for defining the first point). This means that, while defining the profile, an axial straight-line segment needs to be inserted at the first point of the profile (P), to shift the first point up to the axial level of the current tool position. *Effectively, the actual first point on the profile becomes the second point in the definition of the profile.*

On the other hand, the control selects Type-2 if the first point is defined using *both* X/U and Z/W. In this case, the first point on the defined profile need not necessarily be at the current axial position of the tool. The first point may lie to the left (but not to the right) of the current tool position, if the feed motion is towards the chuck, as in Fig. 1. In the case of machining towards the tailstock, the first point may lie to the right (but not to the left) of the current tool position. However, this is unnecessary; the current axial position of the tool and that of the first point on the profile should *preferably* match (for which W0 can be used). This, effectively, means that the actual first point on the profile has been axially shifted to match the current axial position of the tool, requiring insertion of an axial straight-line segment at the actual first point on the profile, as is done in the case of Type-1 machining.

In short, the profile can be defined for both the types in the same manner, with the difference that the presence of W0 in the P block would select Type-2 machining.

Defining the first point on the profile using both X/U and Z/W on a machine, which does not have G71 Type-2 available on it, is not allowed. In such a case, the execution would terminate with the P/S alarm 065.

Permissible Profile for G71 Type-1

G71 Type-1 can produce any type of profile (consisting of linear and circular segments only) which is not having any undercut (i.e., a pocket) anywhere along the axis of the job. In other words, there should be *monotonic* increase or decrease in the diameter, along the axis of the job. For example, if the feed motion is towards the chuck, then the diameter of the job should either increase or remain unchanged in the negative Z direction, in the case of external machining. In internal machining towards the chuck, the diameter should either decrease or remain unchanged in the negative Z direction. For machining towards the tailstock, the variation should be just opposite to what is stated above, in both cases.

The condition of monotonic variation in X is relaxed by G71 Type-2, as discussed later. However, a non-monotonic variation in Z is an impossible profile for G71. In other words, there should not be any pocket when the profile is viewed in the axial direction.

Stock Size for G71

As shown in Fig. 1, G71 adds the specified finishing allowances, with sign, to the defined surface profile (PQ), and considers the new profile thus obtained (CD) the desired profile for machining. S is the position of the tool just before the G71 block, i.e., the current tool position. G71 adds the specified finishing allowances, with sign, to this point to obtain a new point A which it uses for defining the size of the stock. It determines the machining zone by drawing an imaginary radial line from D up to the radial level of A, followed by an axial line to reach A. At the other end, it draws an axial line from C up to the axial level of A (this is because we have axially extended the profile at its first point up to the axial level of the start point S), followed by a radial line to reach A. The material contained in the enclosed area defined by the drawn straight lines and the desired profile CD is removed by G71.

Toolpath of G71 Type-1

G71 Type-1 produces the desired profile by first roughing it and then smoothening it, as shown in Fig. 1, where point P is the axial end of the cylindrical workpiece, and point Q is its radial end. The sequence of motion, which starts and ends at S, is numbered. The feed and the rapid motions are shown by continuous and dashed lines, respectively.

Roughing: Multiple straight-turning operations, at the specified feedrate, up to the desired profile (defined profile PQ plus specified allowances, with sign, shown as CD in this figure as well as in Fig. 2) at successively decreasing radial levels (increasing radial levels, in the case of internal machining) by the specified depth of cut (A to B), producing the desired profile with extra steps which are shown as the shaded triangular shapes in the figure. At the end of each roughing pass, the tool retracts at 45° by the specified radial-retraction distance, at feedrate, followed by a rapid axial retraction to the axial position of A. Thereafter, it moves radially to the start point of the next roughing pass. This motion can be at the rapid rate or feedrate depending on whether G00 or G01 is programmed in the P block of the profile definition (i.e., for defining the first point on the profile). Since this motion does not involve cutting, it should preferably be at the rapid rate, for which the first point on the profile would need to be defined with G00. All other points on the profile (actually, segments of the profile) must be defined using G01/G02/G03 only.

Irrespective of the radial position of D, the roughing starts from the radial position of A (minus depth of cut), up to the desired profile CD or the Z position of D, if a roughing pass happens to lie above D. Therefore, with a large radial clearance, some initial roughing passes will cut in air (Fig. 1 shows three such passes), which should be avoided. In fact, there is no need to give radial clearance. Some axial clearance (say, 2 mm) may be given.

The fact that roughing starts from the radial level of A, can be advantageously used to resume any unfinished G71 because of any reason such as power failure. Repeating the whole cycle all over again would unnecessarily waste time by cutting in air. Instead, the initial tool position (S) may be brought down appropriately (assuming external machining) to eliminate the previously completed roughing passes. Effectively, the cycle resumes from the point where it stopped.

Step removal: With each roughing pass, the depth to be roughed out reduces by the specified depth of cut. The roughing continues till the remaining depth becomes smaller than the specified depth of cut. Thereafter, the tool goes to the radial level of the first point of the desired profile, and exactly follows the profile at feedrate (an axial move to C, followed by C to D, as shown in the figure). This removes the steps created in the roughing operation. After D, it executes a rapid radial move followed by an axial move, to reach the initial tool position, S.

Note that, in the different control versions/modals of Fanuc, there are some minor differences in the toolpath of G71, in the beginning (path 1→2) as well as in the end of the cycle (path 23→24), which may not be the same as what is shown in Fig. 1. For example, the path 1→2 may be combined into a single straight-line path. Similarly, the path 23→24 may show dog-leg effect. However, the syntax of the cycle as well as all the roughing passes and the final step-removal pass are exactly same across the control versions. The programmer need not worry about these insignificant differences which do not affect the actual machining.

Defining the Profile

The profile, which can have both linear and circular segments (only), must be defined from P to Q, i.e., in the direction of the tool movement during the step-removal operation *which proceeds from the last roughing cut towards the first roughing cut.*

Considering right-to-left and left-to-right machining in both external and internal machining, there can be four combinations of acceptable surface profiles. These are shown in Fig. 2, along with the appropriate signs of finishing allowances. The direction of defining the profile is indicated by arrows; P to Q in all cases. The case (d) in this figure is only a theoretical possibility on a common CNC lathe.

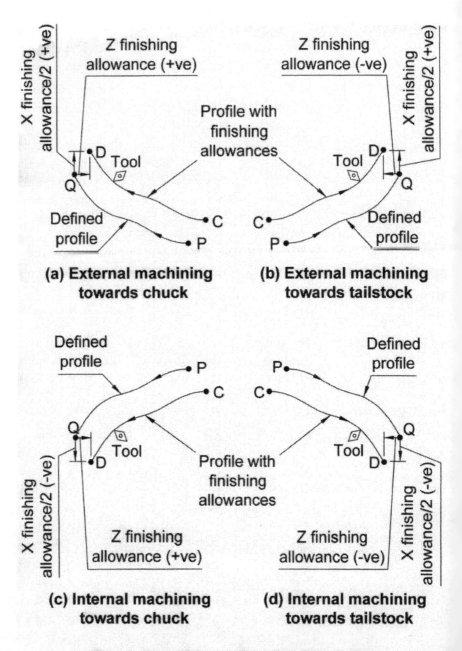

Fig. 2: Possible profiles for G71 Type-1

Selection between G71 and G72

G71 is a very efficient machining operation, provided the extent of the material to be removed is more in the axial direction, compared to that in the radial direction. In case the extent in the axial direction is less than that in the radial direction (e.g., for very small length-to-diameter ratio jobs), the choice of the multiple facing cycle, G72, results in more efficient machining. G72 is described later.

Finishing Cycle G70

After the completion of roughing cycles, the part obtained is slightly over-size, to the extent of the specified finishing allowances. The exact shape can be obtained simply by making the tool move along the profile *PQ*, choosing the appropriate tool and the cutting parameters. This amounts to defining the profile again, after machining by the roughing cycle is over. This is repetitive work because the profile was already defined for the roughing cycle. In order to avoid this repetitive work, a cycle G70 has been designed:

G70 P_ Q_;

The position of the tool in the previous block becomes the start point as well as the end point of G70. Starting from this position, the tool goes to the point defined in the *P* block at the rapid rate or the feedrate, depending on whether the *P* block contains G00 or G01. Thereafter, all the commands up to the *Q* block are sequentially executed, and the tool comes back to the start/end point at the rapid rate, completing the cycle. After this, the execution starts from the block next to the G70 block. One can view G70 as simply a copy/paste operation of the blocks between *P* and *Q*, with additional moves from/to the start/end point.

The *P* and *Q* blocks of both G71 (or G72/G73) and G70 should obviously refer to the same block numbers because the profile is the same for both.

While G00/G01 in the *P* block does not have much effect on the machining time of G71/G72 (because the distance involved is very small), G01 makes G70 slower because the tool would approach the first point on the profile at the feedrate, covering a large distance. G00 avoids this unnecessary feed motion in air. Therefore, as a rule of thumb, the *P* block must use G00.

G70 need not necessarily be used after G71/G72/G73. Though not recommended, one can choose to do away with it by specifying zero finishing allowances in the roughing cycles.

If G70 is desired to be used after G71/G72/G73, even a slightly worn out tool can be used for machining with these roughing cycles. The resulting inaccuracy can be taken care of by specifying suitable finishing allowances which are later removed by G70 with an accurate tool, giving us exact dimensions. This is, in fact, the recommended machining practice.

Sample Program-1

Let us first consider the part shown in Fig. 3 which is required to be produced from a blank size of $\Phi30 \times 70$, by removing the shaded portion by G71 (leaving the specified finishing allowances), followed by G70 (for machining to the exact size). This is an example of external machining towards the chuck.

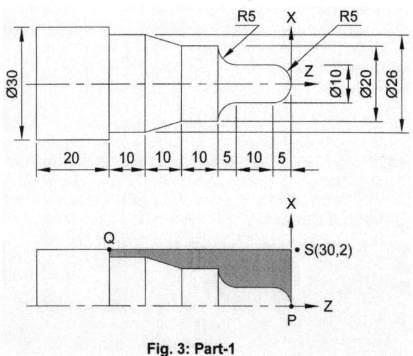

Fig. 3: Part-1

156

In the shaded region, the roughing would start from the top and proceed towards the axis of the workpiece where the last roughing pass would be placed. The step-removal pass, which proceeds from the last roughing cut, towards the first roughing cut, would be from P to Q (with the specified finishing allowances added to these points). Therefore, the profile must be defined from P to Q,

S (30,2) has been chosen as the start/end point of G71. As per the usual practice, no radial clearance but 2 mm axial clearance from the workpiece has been provided. As a consequence of the axial clearance, the definition of the profile would start from (0,2) rather than (0,0), and the actual first point on the profile, P, would actually become the second point.

O0001; (Program number 1)

G21 G96 G99; (mm mode, CSS and feed in mm/rev selected)

G54; (Workpiece coordinate system)

G50 S2000; (The maximum RPM clamped to 2000 for the CSS mode)

G28 U0; (Tool goes to the X reference position)

G28 W0; (Tool goes to the Z reference position. It is a safe practice to send the tool first to the X reference point, and then to the Z reference point, as it eliminates the possibility of any interference during movement. Note that commanding G28 in both the blocks is necessary because it is a non-modal or one-shot G code. G codes belonging to Group 0 are non-modal codes)

T0101; (Tool number 1 and offset number 1 selected for roughing)

G00 Z2;

X30; (Rapid positioning at the start/end point of G71. No radial clearance but 2 mm axial clearance from the workpiece has been provided. An axial move, followed by a radial move, to reach the desired position, eliminates the possibility of any interference during movement. Note that G00 is implied here because it is a modal G code of Group 1. Based on the similarity in their functionality, G codes have been arranged into various Groups such as 0, 1, 2 etc. Except those belonging to Group 0, all G codes

157

are modal codes which remain effective until another G code from the same Group is programmed. At any time, one G code from each group, except Group 0 codes, is effective. Every Group, except Group 0, has a default code which remains effective when the machine is powered ON or when the control is reset. It is, however, a good practice not to rely on defaults, and explicitly program every code that is needed)

M03 S20; (CW spindle starts, maintaining CSS = 20 m/min. In the CSS mode, a large radial change in the rapid mode is not desirable because it unnecessarily loads the spindle which needs to accelerate or decelerate at a high rate to maintain the specified CSS. Therefore, the spindle has been started after placing the tool near the workpiece)

M08; (Coolant starts)

G71 U0.5 R0.2; (Depth of cut 0.5 mm, and radial tool retraction 0.2 mm specified)

G71 P1 Q2 U0.1 W0.1 F0.3; (The profile definition starts at the block with the sequence number N1, and ends at the block with the sequence number N2. Any valid sequence number, other than 1 and 2, can also be used. Except for these two blocks, which G71 searches, sequence numbers for other blocks in the program are optional. The X finishing allowance (on diameter) is 0.1 mm, the Z finishing allowance is 0.1 mm, and the feedrate is 0.3 mm/rev. Both the finishing allowances must be positive for this geometry which resembles case (a) of Fig. 2)

N1 G00 X0; (No Z or W has been specified in this block. Hence, Z2 is implied, and the control would select G71 Type-1 for machining. (0,2) becomes the new first point on the profile. The actual first point (0,0) becomes the second point in the definition of the profile. This block should use G00, rather than G01, because of the reasons explained earlier)

G01 Z0 F0.1; (The block end point is the actual first point P on the profile. The feedrate specified here would be ignored by G71. It would be later used by G70, for this as well as all subsequent blocks. Feedrate is a modal data. It remains effective until it is changed. Therefore, if the next block uses the same feedrate as that in the previous block, then the feedrate need not be commanded in the next block. Also, in any block, any unchanged coordinate need not be specified. Therefore, X0 is implied in this block)

G03 X10 Z–5 R5; (The definition of the profile continues till the block N2. F0.1 is implied)

G01 Z–15;

G02 X20 Z–20 R5;

G01 Z–30;

X26 Z–40; (G01 implied because it is a modal code of G-code Group 1)

Z–50;

N2 X30; (The block end point is the last point Q on the profile)

M05; (Spindle stops. Movement to/from the reference position causes large radial displacement. Therefore, in the CSS mode, it is better to stop the spindle here)

M09; (Coolant stops)

G28 U0; (Tool goes to the X reference position)

G28 W0; (Tool goes to the Z reference position)

T0303; (Tool number 3 and offset number 3 selected for finishing)

G00 Z2;

X30; (Rapid positioning at the start point of G70)

M03; (Spindle started for the finishing cycle G70, at the previously specified CSS. If desired, a different CSS can be specified here)

M08; (Coolant starts)

G70 P1 Q2; (Finishing cycle. It removes the extra material specified as finishing allowances in the second block of G71)

M05; (Spindle stops)

M09; (Coolant stops)

G28 U0; (Tool goes to the X reference position)

G28 W0; (Tool goes to the Z reference position)

M30; (Control resets and the program rewinds)

Sample Program-2

We next consider the part shown in Fig. 4(a) which is required to be produced from a blank size shown in Fig. 4(b), by removing the shaded portion by G71 (leaving the specified finishing allowances), followed by G70 (for machining to the exact size). This is an example of external machining towards the tailstock.

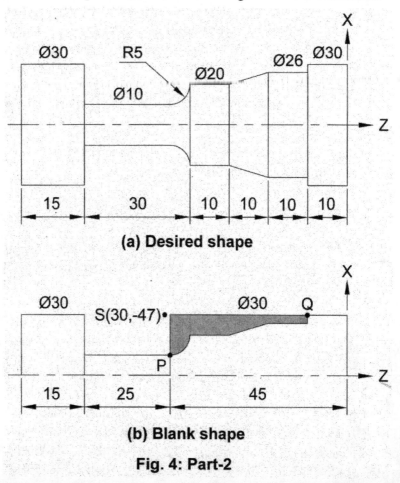

(a) Desired shape

(b) Blank shape

Fig. 4: Part-2

In the shaded region, the roughing would start from the top ($\Phi30$) and proceed downwards towards $\Phi10$ where the last roughing pass would be placed. The step-removal pass, which proceeds from the last roughing cut, towards the first roughing cut, would be from *P* to *Q* (with the specified finishing allowances added to these points). Therefore, the profile must be defined from *P* to *Q*,

S (30,–47) has been chosen as the start/end point of G71. As per the usual practice, no radial clearance but 2 mm axial clearance from the workpiece has been provided. As a consequence of the axial clearance, the definition of the profile would start from (10,–47) rather than (10,–45), and the actual first point on the profile, *P*, would actually become the second point.

O0002; (Program number 2)

G21 G96 G99; (mm mode, CSS and feed in mm/rev selected)

G54; (Workpiece coordinate system)

G50 S2000; (The maximum RPM clamped to 2000 for the CSS mode)

G28 U0; (Tool goes to the *X* reference position)

G28 W0; (Tool goes to the *Z* reference position)

T0505; (Tool number 5 and offset number 5 selected for roughing)

G00 Z–47;

X30; (Rapid positioning at the start/end point of G71. No radial clearance but 2 mm axial clearance from the workpiece has been provided. Note that the axial clearance is towards left because it is left-to-right machining)

M03 S20; (CW spindle starts, maintaining CSS = 20 m/min)

M08; (Coolant starts)

G71 U0.5 R0.2; (Depth of cut 0.5 mm, and radial tool retraction 0.2 mm specified)

G71 P1 Q2 U0.1 W–0.1 F0.3; (The profile definition starts at the block with the sequence number N1, and ends at the block with the sequence number N2. The *X* finishing allowance (on diameter) is 0.1 mm, the *Z* finishing allowance is –0.1 mm, and the feedrate is 0.3 mm/rev. The *X* and *Z* finishing allowances must be positive and negative, respectively, for this geometry which resembles case (b) of Fig. 2)

N1 G00 X10; (No Z or W has been specified in this block. Hence, Z–47 is implied, and the control would select G71 Type-1 for machining. (10,–47) becomes the new first point on the profile. The actual first point (10,–45) becomes the second point in the definition of the profile)

G01 Z–45 F0.1; (The block end point is the actual first point P on the profile)

G03 X20 Z–40 R5; (The definition of the profile continues till the block N2)

G01 Z–30;

X26 Z–20;

Z–10;

N2 X30; (The block end point is the last point Q on the profile)

M05; (Spindle stops)

M09; (Coolant stops)

G28 U0; (Tool goes to the X reference position)

G28 W0; (Tool goes to the Z reference position)

T0707; (Tool number 7 and offset number 7 selected for finishing)

G00 Z–47;

X30; (Rapid positioning at the start point of G70)

M03; (Spindle started for the finishing cycle G70)

M08; (Coolant starts)

G70 P1 Q2; (Finishing cycle)

M05; (Spindle stops)

M09; (Coolant stops)

G28 U0; (Tool goes to the X reference position)

G28 W0; (Tool goes to the Z reference position)

M30; (Control resets and the program rewinds)

Sample Program-3

We next consider the part shown in Fig. 5(a) which is required to be produced from a blank size (a hollow cylinder) shown in Fig. 5(b), by removing the shaded portion by G71 (leaving the specified finishing allowances), followed by G70 (for machining to the exact size). This is an example of internal machining towards the chuck.

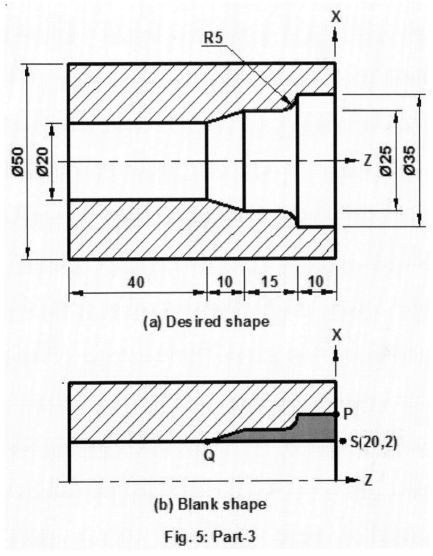

(a) Desired shape

(b) Blank shape

Fig. 5: Part-3

In the shaded region, the roughing would start from the inner surface ($\Phi20$) and proceed upwards towards $\Phi35$ where the last roughing pass would be placed. The step-removal pass, which proceeds from the last roughing cut, towards the first roughing cut, would be from P to Q (with the specified finishing allowances added to these points). Therefore, the profile must be defined from P to Q,

S (20,2) has been chosen as the start/end point of G71. As per the usual practice, no radial clearance but 2 mm axial clearance from the workpiece has been provided. As a consequence of the axial clearance, the definition of the profile would start from (35,2) rather than (35,0), and the actual first point on the profile, P, would actually become the second point.

O0003; (Program number 3)

G21 G96 G99; (mm mode, CSS and feed in mm/rev selected)

G54; (Workpiece coordinate system)

G50 S2000; (The maximum RPM clamped to 2000 for the CSS mode)

G28 U0; (Tool goes to the X reference position)

G28 W0; (Tool goes to the Z reference position)

T0202; (Tool number 2 and offset number 2 selected for roughing)

G00 Z2;

X20; (Rapid positioning at the start/end point of G71. No radial clearance but 2 mm axial clearance from the workpiece has been provided)

M03 S20; (CW spindle starts, maintaining CSS = 20 m/min)

M08; (Coolant starts)

G71 U0.5 R0.2; (Depth of cut 0.5 mm, and radial tool retraction 0.2 mm specified)

G71 P1 Q2 U−0.1 W0.1 F0.3; (The profile definition starts at the block with the sequence number N1, and ends at the block with the sequence number N2. The X finishing allowance (on diameter) is −0.1 mm, the Z finishing allowance is 0.1 mm, and the feedrate is 0.3 mm/rev. The X and Z finishing allowances must be negative and positive, respectively, for this geometry which resembles case (c) of Fig. 2)

N1 G00 X35; (No *Z* or *W* has been specified in this block. Hence, *Z2* is implied, and the control would select G71 Type-1 for machining. (35,2) becomes the new first point on the profile. The actual first point (35,0) becomes the second point in the definition of the profile)

G01 Z0 F0.1; (The block end point is the actual first point *P* on the profile)

Z−10; (The previous block and this block can be combined because it is the same straight line)

G02 X25 Z−15 R5; (The definition of the profile continues till the block N2)

G01 Z−25;

N2 X20 Z−35; (The block end point is the last point *Q* on the profile)

M05; (Spindle stops)

M09; (Coolant stops)

G28 U0; (Tool goes to the *X* reference position)

G28 W0; (Tool goes to the *Z* reference position)

T0404; (Tool number 4 and offset number 4 selected for finishing)

G00 Z2;

X20; (Rapid positioning at the start point of G70)

M03; (Spindle started for the finishing cycle G70)

M08; (Coolant starts)

G70 P1 Q2; (Finishing cycle)

M05; (Spindle stops)

M09; (Coolant stops)

G28 U0; (Tool goes to the *X* reference position)

G28 W0; (Tool goes to the *Z* reference position)

M30; (Control resets and the program rewinds)

Radius Compensation in Roughing Cycles

The earlier control versions do not incorporate radius compensation in G71/G72/G73. Even if G41/G42 is active, it is simply ignored by these roughing cycles. Machining with radius compensation is possible only at the time of the final machining with the finishing cycle G70, which can be done in two ways:

(1) Commanding G41/G42 in the *P* block of G71/G72/G73

(2) Commanding G41/G42 just before the G70 block

While both the methods work the same way, it is better to use the second method because the first method gives the misleading impression that the roughing cycle is also using radius compensation.

As an example of the second method in the first program, the G70 block, along with the preceding rapid-positioning move (X30), can be replaced by

G42 X30;

M03 M08;

G70 P1 Q2;

G40 X35 Z5;

Explicitly canceling the compensation mode using G40 is not needed in this program. G28 automatically cancels it. It is, however, a good practice to cancel compensation after its use is over. Note that the compensation canceled by G28 is again automatically restored if there is a G00/G01 block after G28 (Any other move command, immediately following G28, would alarm out).

Also note that there should not be two or more consecutive non-movement commands in the radius compensation mode; otherwise, there would be some error in compensation at the start point of the following block. This means that the two M codes before G70 would need to be combined in a single block, as done here. If the presence of more than one M code in a block alarms out, set parameter 3404#7 (M3B) = 1 which would allow up to three M codes in one block.

In the case of internal machining, the center of the nose is required to shift towards the left of the programmed path (*P* to *Q*). This would need G41. For example, the third program would need

G41 X20;

M03 M08;

G70 P1 Q2;

G40 X25 Z5;

If the control is capable of incorporating radius compensation in roughing cycles also, the first method should be used. One can also choose to command G41/G42 before the roughing-cycle block.

Note that even if the control does not use radius compensation in roughing cycles, it is not a limitation, as regards the accuracy, provided the center of the nose is not selected as the reference point of the tool (which nobody does). For example, in the case of external machining towards the chuck, the position of the reference point, in the usual method of offset setting, corresponds to nose number 3 (the imaginary lower left extreme of the external tool). If machining is done with this reference point, without using radius compensation, then there would be no error in straight turning (because the *X* position of the reference point and that of the bottom-most point of the tool, which is the cutting point in straight turning, are same) as well as in straight facing (because the *Z* position of the reference point and that of the left-most point of the tool, which is the cutting point in straight facing, are same). There would, of course, be some error in taper turning/facing as well as in circular motion (i.e., whenever there is *simultaneous* movement along both the *X* and the *Z* axes) but the resulting part would only be slightly over-size, if machined without radius compensation. This can be machined to the exact size by using G70 with radius compensation. And, if the finishing tool has a very small nose radius, it might give acceptable accuracy even without radius compensation in most cases.

Actually, the roughing cycles are never meant for very accurate machining. We use large depth of cut and high feedrate with a large nose-radius tool, with an intention to reduce the cycle time. Finishing allowances are provided to take care of the resulting inaccuracy and surface roughness. It is G70 which gives the exact dimensions with good surface finish. Therefore, it hardly matters whether or not the roughing cycles incorporate radius compensation.

G71 Type-2 Toolpath

G71 Type-2 allows a non-monotonic profile also between the P and the Q blocks, i.e., a profile with pocket(s) on the surface of the job is permitted. The roughing is similar to the Type-1 toolpath with two major differences:

(1) The retraction at the end of a roughing pass is along the profile, which does not leave any step in roughing.

(2) Roughing is done in pockets also where the tool enters the pocket along the profile and retracts along the other side of the profile, after roughing (straight turning). Again, no steps are left.

The roughing is initially done ignoring the pockets. After it is complete, the pockets are machined one-by-one. Nesting of pockets, i.e., another pocket in a pocket is also permitted. There are some differences in the toolpath on different control versions, but all give the same final shape, and programmed in the same manner. Therefore, there is no need to worry about the exact sequence of the toolpath. As far as programming is concerned, the method is exactly same as that in Type-1, with the difference that Type-2 will have both X/U and Z/W addresses in the P block.

An important point to note is that the Z finishing allowance for a profile with pocket(s) must be zero; otherwise, while one side of the pocket would be left with some extra material specified as finishing allowance, the opposite side would be overcut, spoiling the part. It is not possible to specify finishing allowances with opposite signs for the opposite sides of a pocket. In such cases, only X finishing allowance can be specified; positive in external machining, and negative in internal machining.

Another important point is the selection of an appropriate tool for machining a pocket. The cycle only ensures that the reference point of the tool moves along the desired profile. It cannot check for any interference with the body of the tool. In

168

most cases, only one side of a pocket would be made correctly; the other side would be gouged by the trailing edge of the insert. This can often be avoided by using a neutral tool with small tip angle. If the cycle supports radius compensation, then a small-radius round insert would be the most appropriate, which can make even a semicircular pocket.

MULTIPLE ROUGH-FACING CYCLE, G72

Introduction

This cycle is similar to the multiple rough-turning cycle, G71, with the difference that the roughing is done radially in this cycle. The syntax also is the same except that the depth of cut in G72 is designated by W rather than U in the first block.

Toolpath of G72 Type-1

The toolpath of G72 Type-1 is shown in Fig. 6. Since the descriptions of the two cycles are similar, it is not being repeated for G72. Instead, only the salient points are briefly mentioned below.

Choosing the Start Point

The start point of the cycle should be selected so as to give about 2 mm radial clearance but no axial clearance with the workpiece. Axial clearance results in some wasteful initial roughing passes in air, as shown in Fig. 6.

Defining the Profile

The step-removal pass is from the last roughing pass to the first roughing pass. The profile is required to be defined this way only, from P to Q in this figure. Thus, for the same job, the direction of defining the profile is opposite to that for G71.

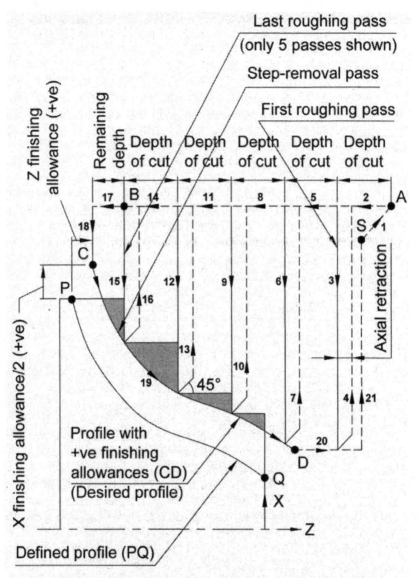

Fig. 6: G72 Type-1 toolpath

Note: S is the position of the tool just before the G72 block, i.e., the current tool position. This point automatically becomes the start/end point of G72.

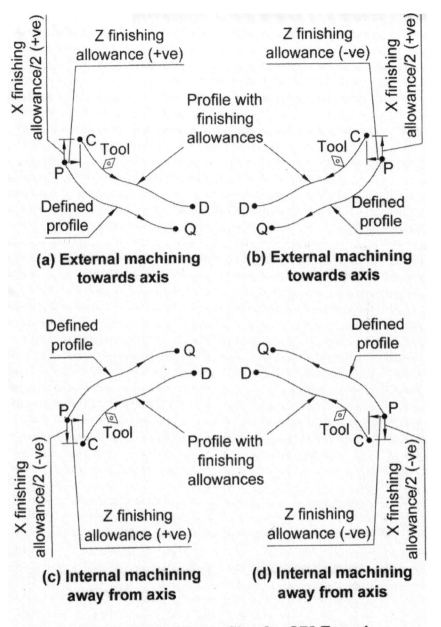

Fig. 7: Possible profiles for G72 Type-1

Finishing Allowances

The logic implemented by the control is independent of G71/G72/G73. In all the three cases, the specified finishing allowances, with sign, are added to the defined profile to obtain the desired profile. Thus, Fig. 2, which shows the effect of the finishing allowances in different cases, is valid for G72 also. Figure 7 is the modified version of Fig. 2, to suit G72. The profile needs to be defined from P to Q in all the four cases of this figure where case (d) is only a theoretical possibility.

Permissible Profile for G72

The Type-1 cycle does not allow a non-monotonous profile. There should not be any pocket when the profile is viewed in the axial direction. In other words, as the profile is defined from the first point (P) to the last point (Q), the Z coordinate should either always increase (or remain constant) or always decrease (or remain constant). Type-2 relaxes this condition by allowing a non-monotonic variation in Z. However, a non-monotonic profile in X is an impossible profile for G72, which alarms out.

Selection between G72 Type-1 and Type-2

The control selects Type-1 if only Z or W is commanded in the P block. If both X/U and Z/W are commanded then Type-2 is selected. Specifying the both would alarm out on a machine on which Type-2 is not available.

If X/U is not specified in the P block, then the radial position of the first point is assumed to be the same as that of the start/end point of the cycle (S in Fig. 6). Effectively, the profile gets radially extended at its actual first point, P, up to the radial level of S. Consequently, the point P becomes the second point while defining the profile.

Stock Size for G72

The stock size for G72 is determined in a similar way it is done for G71. Referring to Fig. 6, point A is located by adding the specified finishing allowances to the start/end point S. The machining zone is determined by drawing an imaginary axial line from D up to the axial level of A, followed by a radial line to reach A. At the other end, a radial line from C is drawn up to the radial level of A (this is because the profile has been radially extended at its first point up to the radial level of the start point S), followed by an axial line to reach *A*. The material contained in the enclosed area defined by the drawn straight lines and the desired profile *CD* is removed by G72.

The axial extents of the machining zones of the jobs (the shaded areas) shown in Figs. 1, 2 and 3 are larger than their radial extents. Therefore, while these can be machined by both G71 as well as G72, machining with G71 would be more efficient. However, these are being taken as sample examples for G72 also, with the purpose to illustrate the differences in the programming methodologies of the two cycles for the same jobs.

Sample Program-4

This refers to the job shown in Fig. 3.

O0004; (Program number 4)

G21 G96 G99; (mm mode, CSS and feed in mm/rev selected)

G54; (Workpiece coordinate system)

G50 S2000; (The maximum RPM clamped to 2000 for the CSS mode)

G28 U0; (Tool goes to the *X* reference position)

G28 W0; (Tool goes to the *Z* reference position)

T0101; (Tool number 1 and offset number 1 selected for roughing)

G00 Z0;

X34; (Rapid positioning at the start/end point of G72. No axial clearance but 2 mm radial clearance from the workpiece has been provided)

M03 S20; (CW spindle starts, maintaining CSS = 20 m/min)

M08; (Coolant starts)

G72 W0.5 R0.2; (Depth of cut 0.5 mm, and radial tool retraction 0.2 mm specified)

G72 P1 Q2 U0.1 W0.1 F0.3; (The profile definition starts at the block with the sequence number N1, and ends at the block with the sequence number N2. The *X* finishing allowance (on diameter) is 0.1 mm, the *Z* finishing allowance is 0.1 mm, and the feedrate is

0.3 mm/rev. Both the finishing allowances must be positive for this geometry which resembles case (a) of Fig. 7)

N1 G00 Z–50; (No *X* or *U* has been specified in this block. Hence, X34 is implied, and the control would select G72 Type-1 for machining. (34, –50) becomes the new first point on the profile. The actual first point (30,–50) becomes the second point in the definition of the profile. This block should use G00, rather than G01, because of the reasons explained earlier)

G01 X30 F0.1; (The block end point is the actual first point *Q* on the Fig. 3 profile. The feedrate specified here would be ignored by G72. It would be later used by G70, for this as well as all subsequent blocks)

X26; (This can be combined with the previous block)

Z–40;

X20 Z–30;

Z–20;

G03 X10 Z–15 R5;

G01 Z–5;

N2 G02 X0 Z0 R5; (The block end point is the last point *P* on the Fig. 3 profile)

M05; (Spindle stops)

M09; (Coolant stops)

G28 U0; (Tool goes to the *X* reference position)

G28 W0; (Tool goes to the *Z* reference position)

T0303; (Tool number 3 and offset number 3 selected for finishing)

G00 Z0;

X34; (Rapid positioning at the start point of G70)

M03; (Spindle started for the finishing cycle G70, at the previously specified CSS)

M08; (Coolant starts)

G70 P1 Q2; (Finishing cycle. It removes the extra material specified as finishing allowances in the second block of G72)

M05; (Spindle stops)

M09; (Coolant stops)

G28 U0; (Tool goes to the *X* reference position)

G28 W0; (Tool goes to the *Z* reference position)

M30; (Control resets and the program rewinds)

Sample Program-5

This refers to the job shown in Fig. 4.

O0005; (Program number 5)

G21 G96 G99; (mm mode, CSS and feed in mm/rev selected)

G54; (Workpiece coordinate system)

G50 S2000; (The maximum RPM clamped to 2000 for the CSS mode)

G28 U0; (Tool goes to the *X* reference position)

G28 W0; (Tool goes to the *Z* reference position)

T0505; (Tool number 5 and offset number 5 selected for roughing)

G00 Z−45;

X34; (Rapid positioning at the start/end point of G72. No axial clearance but 2 mm radial clearance from the workpiece has been provided)

M03 S20; (CW spindle starts, maintaining CSS = 20 m/min)

M08; (Coolant starts)

G72 W0.5 R0.2; (Depth of cut 0.5 mm, and radial tool retraction 0.2 mm specified)

G72 P1 Q2 U0.1 W−0.1 F0.3; (The profile definition starts at the block with the sequence number N1, and ends at the block with the sequence number N2. The *X* finishing allowance (on diameter) is 0.1 mm, the *Z* finishing allowance is −0.1 mm, and the feedrate is 0.3 mm/rev. The *X* and *Z* finishing allowances must be positive and negative, respectively, for this geometry which resembles case (b) of Fig. 7)

N1 G00 Z–10; (No *X* or *U* has been specified in this block. Hence, X34 is implied, and the control would select G72 Type-1 for machining. (34,–10) becomes the new first point on the profile. The actual first point (30,–10) becomes the second point in the definition of the profile)

G01 X30 F0.1; (The block end point is the actual first point *Q* on the Fig. 4 profile)

X26; (This can be combined with the previous block)

Z–20;

X20 Z–30;

Z-40;

N2 G02 X10 Z–45 R5; (The block end point is the last point *P* on the Fig. 4 profile)

M05; (Spindle stops)

M09; (Coolant stops)

G28 U0; (Tool goes to the *X* reference position)

G28 W0; (Tool goes to the *Z* reference position)

T0707; (Tool number 7 and offset number 7 selected for finishing)

G00 Z–45;

X34; (Rapid positioning at the start point of G70)

M03; (Spindle started for the finishing cycle G70)

M08; (Coolant starts)

G70 P1 Q2; (Finishing cycle)

M05; (Spindle stops)

M09; (Coolant stops)

G28 U0; (Tool goes to the *X* reference position)

G28 W0; (Tool goes to the *Z* reference position)

M30; (Control resets and the program rewinds)

Sample Program-6

This refers to the job shown in Fig. 5.

O0006; (Program number 6)

G21 G96 G99; (mm mode, CSS and feed in mm/rev selected)

G54; (Workpiece coordinate system)

G50 S2000; (The maximum RPM clamped to 2000 for the CSS mode)

G28 U0; (Tool goes to the *X* reference position)

G28 W0; (Tool goes to the *Z* reference position)

T0202; (Tool number 2 and offset number 2 selected for roughing)

G00 Z5;

X16;

Z0; (Rapid positioning at the start/end point of G72. No axial clearance but 2 mm radial clearance from the workpiece has been provided. Note that the radial clearance is in the negative *X* direction because the feed motion is in the positive *X* direction)

M03 S20; (CW spindle starts, maintaining CSS = 20 m/min)

M08; (Coolant starts)

G72 W0.5 R0.2; (Depth of cut 0.5 mm, and radial tool retraction 0.2 mm specified)

G72 P1 Q2 U–0.1 W0.1 F0.3; (The profile definition starts at the block with the sequence number N1, and ends at the block with the sequence number N2. The *X* finishing allowance (on diameter) is –0.1 mm, the *Z* finishing allowance is 0.1 mm, and the feedrate is 0.3 mm/rev. The *X* and *Z* finishing allowances must be negative and positive, respectively, for this geometry which resembles case (c) of Fig. 7)

N1 G00 Z–35; (No *X* or *W* has been specified in this block. Hence, X16 is implied, and the control would select G72 Type-1 for machining. (16, –35) becomes the new first point on the profile. The actual first point (20,–35) becomes the second point in the definition of the profile)

G01 X20 F0.1; (The block end point is the actual first point *Q* on the Fig. 5 profile)

X25 Z–25;

Z–15;

G03 X35 Z–10 R5;

N2 G01 Z0; (The block end point is the last point *P* on the Fig. 5 profile)

M05; (Spindle stops)

M09; (Coolant stops)

G00 Z5; (Tool retracted to avoid interference)

G28 U0; (Tool goes to the *X* reference position)

G28 W0; (Tool goes to the *Z* reference position)

T0404; (Tool number 4 and offset number 4 selected for finishing)

G00 Z5;

X16;

Z0; (Rapid positioning at the start point of G70)

M03; (Spindle started for the finishing cycle G70)

M08; (Coolant starts)

G70 P1 Q2; (Finishing cycle)

M05; (Spindle stops)

M09; (Coolant stops)

G00 Z5; (Tool retracted to avoid interference)

G28 U0; (Tool goes to the *X* reference position)

G28 W0; (Tool goes to the *Z* reference position)

M30; (Control resets and the program rewinds)

POCKET MACHINING WITH TYPE-1 CYCLES

Actually, it is not possible. But, there is a way: First, define the profile ignoring the pockets. In other words, eliminate the pocket(s) in the profile by drawing axial or radial lines, as appropriate, to cover the pocket(s). The resulting profile would be monotonous which can be machined by G71/G72 Type-1 cycle. We will end up with a part which will have all the features except the pocket(s). These pockets can be machined by the Pattern Repeating Cycle, G73, discussed earlier (Volume 2). G73 was originally designed for machining cast/forged parts, but it can also be very effectively used for producing pockets.

Note that one can choose to machine the entire workpiece solely by G73 also, but it would be a highly inefficient machining method. G73 is too slow a process. It has to be used judiciously, only when its use cannot be avoided. As a rule of thumb, G71/G72 must be employed to the extent possible. Once the overall shape of the job is obtained, the remaining features can be produced by other methods.

Made in the USA
Las Vegas, NV
31 January 2024

85132246R00101